Land for Industrial Development

OTHER TITLES FROM E & FN SPON

European Directory of Property Developers, Investors and Financiers 1993
2nd edition
Building Economics Bureau

The Idea of Building
S. Groák

Investment Appraisal and Financial Decisions
4th Edition
S. Lumby

Project Management Demystified
Today's tools and techniques
G. Reiss

Microcomputers in Property
A surveyor's guide to Lotus 1-2-3 and dBase IV
T.J. Dixon, O. Bevan and S. Hargitay

The Multilingual Dictionary of Real Estate
L. van Breughel, R.H. Williams and B. Wood

National Taxation for Property Management and Valuation
A. MacLeary

Property Investment Decisions
S. Hargitay and M. Yu

Property Development
3rd Edition
D. Cadman and L. Austin-Crowe
Edited by R. Topping and M. Avis

Property Investment and the Capital Markets
G.R. Brown

Property Valuation
The five methods
D. Scarrett

Residential Estimator
Marshall and Swift

Risk Analysis in Project Management
J. Raftery

Spon's Budget Estimating Handbook
B.S. Spain

Risk, Uncertainty and Decision-making in Property Development
P.J. Bryne and D. Cadman

Transport, the Environment and Sustainable Development
D. Banister and K. Button

UK Directory of Property Developers, Investors and Financiers 1993
7th edition
Building Economics Bureau

Urban Regeneration
Property investment and development
J. Berry, W. Deddis and W. McGreal

Effective Writing
Improving scientific, technical and business communication
2nd Edition
C. Turk and J. Kirkman

For more information on these and other titles please contact:
The Promotion Department, E & FN Spon, 2-6 Boundary Row,
London SE1 8HN Telephone 071-865 0066

Land for Industrial Development

David Adams
University of Aberdeen, UK

Lynne Russell
University of Manchester, UK

and

Clare Taylor-Russell
Trafford Metropolitan Borough Council, UK

Published by E & FN Spon, an imprint of Chapman & Hall, 2-6 Boundary Row, London SE1 8HN, UK

Chapman & Hall, 2-6 Boundary Row, London SE1 8HN, UK

Blackie Academic & Professional, Wester Cleddens Road, Bishopbriggs, Glasgow G64 2NZ, UK

Chapman & Hall GmbH, Pappelallee 3, 69469 Weinheim, Germany

Chapman & Hall USA, One Penn Plaza, 41st Floor, New York, NY 10119, USA

Chapman & Hall Japan, ITP-Japan, Kyowa Building, 3F, 2-2-1 Hirakawa-cho, Chiyoda-ku, Tokyo 102, Japan

Chapman & Hall Australia, Thomas Nelson Australia, 102 Dodds Street, South Melbourne, Victoria 3205, Australia

Chapman & Hall India, R. Seshadri, 32 Second Main Road, CIT East, Madras 600 035, India

First edition 1994

Printed in Great Britain at the University Press, Cambridge

ISBN 0 419 19180 1

A catalogue record for this book is available from the British Library

Printed on permanent acid-free text paper,manufactured in accordance with ANSI/NISO Z39.48-1992 and ANSI/NISO Z39.48-1984 (Permanence of Paper)

To Judith, Tom, Roman, Catherine and Johanna

CONTENTS

PREFACE

Land is fundamental to all economic activity. The supply of employment land and property is central to the pattern and process of economic regeneration and restructuring, but how these markets operate is poorly understood. The evidence clearly belies the traditional assumption that markets work perfectly, that supply responds to demand, that the price mechanism adjusts to provide an equilibrium at which the needs of industry and commerce are met and available land is taken up. Such a model struggles to reconcile land vacancy and urban dereliction in some areas with overheating in land and property markets in other areas such as the south-east.

Although current economic and social policy is driven by an emphasis on the role of market forces, constraints on land supply and development suggests that public intervention is necessary, either at a local level to ensure an adequate supply of employment land or at a national level to redress regional inequalities in economic performance. It is only relatively recently, however, that land and property markets have become the subject of theoretical discussion. As a result, there is little empirical evidence on the key processes of land exchange and development on which to base any recommendations for public policy or any evaluation of existing policy instruments.

We hope that this volume will go some way to redress that particular imbalance. The discussion and conclusions focus on the supply of land for industrial development and draw on evidence from a detailed study of a local land market area in the north-west of England. A total of 120 sites with potential for industrial development were examined and each of these is the subject of a separate case study. These are used in the text, not in full, but to highlight the behaviour of particular actors in the land market, the impact of development constraints and the mechanisms which successfully bring land forward for development.

The characteristics of the study area, and the contrasts within it (the decline of traditional manufacturing and the growth of high technology and service industries; planning restraint in some areas, planning flexibility in others; inner urban areas and smaller free standing towns; and a variety of public policy instruments) make it an ideal environment within which to consider the processes, constraints and agents which influence land supply, and also suggest a wider relevance for its conclusions.

ACKNOWLEDGEMENTS

The research which forms the basis of this book was undertaken in the Department of Planning and Landscape at the University of Manchester. The authors would like to thank colleagues there, together with all members of the Research Advisory Group, for their support during that time. Members of the Research Advisory Group were: Howard Allen, Marie Bintley, Roger Bristow, Peter Hampson, Tim Harrison, Norman Lee, Paul McNamara, Jim Shaw, Paul Syms.

The research would not have been possible without the considerable cooperation of all those who agreed to be involved. These included planning officers and economic development officers at Cheshire County Council and four district councils — Chester, Crewe & Nantwich, Ellesmere Port & Neston and Wirral; district officers of the Inland Revenue Valuation Office; key local property agents; and senior personnel in 120 manufacturing companies and development companies. All gave generously of their time.

A special word of thanks must go to Jane Preston Associates for the design and production, in camera ready form, of this final text; to Mary Howcroft and Meg Woolley for secretarial support during the project; and to Margaret Bryant and Tim Pope for research assistance.

This work is based on research funded by the Economic and Social Research Council (Award No. R000231427). The support of the Council is gratefully acknowledged. Some of the material in Chapters 1, 3 and 4 has previously been published in *Regional Studies*, *Land Use Policy* and the *Journal of Property Research*.

PART ONE

THEORETICAL DEBATES AND DATA SOURCES

1

LAND MARKETS AND THE NEEDS OF INDUSTRY: A THEORETICAL OVERVIEW

Of all the factors of production, land has been perhaps the least studied empirically. In contrast, a rich diversity of investigative work has been undertaken since the 1970s on the changing relationship of labour and capital in the production process (Massey and Meegan, 1982; Lloyd and Shutt, 1983). This has helped enlighten the wider debate about specific aspects of industrial change such as the apparent growth of new and small firms (O'Farrell and Hitchens, 1988; Mason, 1989), the spatial concentration of high technology industry (Breheny and McQuaid, 1985; Oakey and Cooper, 1989) and the global restructuring of production (Henderson and Castells, 1987). This expanding and detailed literature is informed, however, by only a very limited understanding of how the important processes of land supply, land exchange and land development contribute to meeting the production requirements of industry at the local level. Although the analysis of land and property markets sits at the fringes of both geography and economics, both disciplines have implicitly assumed that there is a smooth and automatic adjustment in supply to meet demand. Until recently neither discipline had challenged this assumption with detailed theoretical or empirical analysis.

This chapter will highlight the most salient points of recent theoretical debate on the operation of the land market. It will be shown that the neo-classical approach, which assumed perfect competition and profit maximization as the basis for industrial location decisions, has become tempered by an acknowledgement of the influence of nonpecuniary preferences and uncertainty by decision makers. On the supply side it has increasingly been recognized that demand, and its influence through the price mechanism, will not necessarily bring forward the land which industry may need. The motives for land ownership are several, the factors which influence supply are complex and an equilibrium may not be achieved — indeed the industrial land market cannot be viewed as a single entity but an interconnected series of submarkets. These refinements in the theoretical approach to the land market have involved discussion of the imperfections and constraints which influence market processes and outcomes. In particular, the literature has recently given attention to the role of the planning system, physical and infrastructural site characteristics, land ownership and valuation methods as impediments to the smooth operation of the land market. The debate has also begun to focus on the influence of such constraints on the various agencies and actors involved in the development process.

The shortage of empirical investigation into the practical operation of land and property markets in the United Kingdom makes it extremely difficult to assess analytically how well markets cater for the space requirements of industry. The extent to which land and property markets influence, or even confine, the locational choices presented to manufacturers is not fully understood, but it is evident that the supply of land may be restricted by particular constraints. The severity of these constraints in different areas and the ways in which they may be overcome have been related to the processes of economic restructuring and in particular to the decentralization of manufacturing industry, or the 'urban—rural shift' (Fothergill, Monk and Perry, 1987). It will be important for the formulation of appropriate policies to ensure an adequate supply of land for industrial development that there is a better understanding of land exchange and development, and of the factors which constrain these processes.

A major study of industrial sites in the north-west of England has afforded an opportunity to explore these issues in some detail, drawing on evidence from 120 case studies. The geographic area for study was chosen in such a way as to represent a local land market with characteristics which would be replicated elsewhere in the country and the discussion will therefore have wider relevance to practitioners and policy makers alike.

1.1 LAND MARKETS AND INDUSTRIAL LOCATION

The empirical neglect of land and property markets has been encouraged by conventional theoretical perspectives which assume that these markets function in a perfectly competitive manner. Indeed, Evans (1985) notes:

> *As urban economics has developed over the last 20 years or so the assumption has almost invariably been made that the land market works smoothly and efficiently to ensure that each parcel of land is used by the activity which can pay the highest rent. Any possible*

> *departures from this norm have usually been ignored in theoretical analysis...* (p. 133).

Underpinning this model are the assumptions that consumers act in a rational manner in economic terms, that there is perfect information and that supply automatically responds to meet expressed demand. More recent theoretical developments have suggested alternative approaches to industrial location decisions and to land supply.

1.1.1 Industrial location theory and the demand for land and property

Conventional industrial location theory derived from Weber (1929) and developed by Losch (1954) and Isard (1956), has assumed that firms behave in a rational and logical manner and are able to achieve their optimal location. Rational behaviour in location decision making was felt to be determined by the aim of profit maximization. While Weber considered that the least cost site was most likely to enable profits to be maximized, Losch preferred the location where demand was greatest. Isard introduced substitution analysis which allowed additional resource inputs to be considered within the structure of Weber's basic proposition. Such neo-classical models depend upon the existence of perfect competition. When the conditions of perfect competition break down, neo-classical models of this type are unlikely to prove helpful in explaining industrial location decisions. Oligopolistic producers, for instance, may be as concerned to maintain market share as to maximize profits.

Subsequently Smith (1966; 1981) has theorized that firms will be prepared to locate at any one of a number of suboptimal points which lie within the *'spatial margins to profitability'*. This is because firms are unable to predict future events with certainty and are therefore unable to undertake all necessary calculations for all possible locations. The spatial margins are defined by the points at which total cost equals total revenue. The area covered by the spatial margins to profitability can be quite extensive, according to the limited empirical evidence available (Taylor, 1970; McDermott, 1973). The availability of information at the time of the decision is therefore likely to be critical in determining where firms actually locate within the spatial margins.

Behavioural approaches, however, concentrate not necessarily on profits, but on the priorities and characteristics of the particular decision makers as determinants of industrial location. An individual entrepreneur may well have a very different agenda in locational decision making than the managing director of a major public company. Within companies, production managers may seek locations which minimize costs while sales managers are likely to prefer those which maximize revenue. Prestige, stability and psychic income, derived from social, environmental and other nonmonetary factors, may also be important. Watts (1987) suggests that locational decisions may even be determined by convenience or proximity to the owner's home.

Townroe (1991) contrasts decisions based on instrumental rationality, which seek the optimal location, with those based on procedural or expressive rationality. The latter may reflect social, historical and cultural factors and thus seek '*the right location*', in the context of what is and is not possible. As Townroe states: *'One feature of industrial location decisions that is very well established empirically is that the elements of risk and uncertainty are typically high, including uncertainty by the decision-maker as to the appropriate procedure of decision-making itself'* (pp. 389-90). Locational behaviour in such an uncertain environment can be seen as 'satisficing' in character, in that entrepreneurs seek satisfactory rather than optimal solutions. A satisfactory location will yield the level of profit which entrepreneurs can reasonably expect to achieve, given their knowledge and abilities at the time of the decision.

More recent theories of industrial location have thus abandoned the pretence of the optimal location and have acknowledged that locational decisions are often surrounded by mists of uncertainty and personal preference. It could also be added that once the assumption of smooth and efficient land markets is relaxed, then the optimal location, even if known, may be unachievable, irrespective of the price offered.

1.1.2 Land supply

Most locational theory pays insufficient attention to the processes through which land and property in fact become available to potential users. Land and property markets are not necessarily compliant. Locational choices can be constrained by market conditions. Indeed, the restricted availability of sites and premises may well be a critical, if insufficiently acknowledged, determinant of industrial location. According to Evans (1985) there are a number of reasons why land may be prevented from passing to the highest possible bidder. In practice, these undermine the conventional theoretical view of the operation of the land market. Evans draws attention to owners' expectations and behaviour, to the cost of redevelopment and upheaval and particularly to the existence of planning controls. Each of these will now be examined in turn.

(a) Owner expectations and behaviour

Theoretical understandings of the land market have begun to widen to consider whether and how the individual preferences of landowners, and in particular their relative appreciation of welfare and substitution effects, might influence the overall supply of land. Within the recent theoretical debate involving Evans (1983; 1986), Wiltshaw (1985; 1988) and Neutze (1987), attention has been given to the extent to which the individual preferences of particular landowners can be accommodated in theoretical models of land supply. Evans, for instance, acknowledges that the reluctance of elderly couples to sell up and move at any price, or the high regard with which wealthy landowners may hold the amenity of their estates over and above any tempting offers received from developers, could be the kinds of exceptional cases where individual preferences are more important than monetary considerations. Wiltshaw (1985) discusses the extent to which land affords the owner satisfaction in a non-market use (for instance as a garden) and subsequently (1988) explains the extent to which non-land income is likely to affect supply prices. As soon as theory begins to take account of individual preferences in this way, it becomes much more difficult to construct an aggregate supply curve. Indeed, the entire debate has demonstrated that *'a considerable amount of work needs to be done before the economic theory of the land market can be considered to be on a par with the economics of other factors of production'* (Evans, 1986, p. 530).

Goodchild and Munton (1985) have identified at least four motives for land ownership. As well as occupation, investment and making land available for others to use on a non-profit basis, some owners may hold land purely for the purpose of control. Such owners may wish to restrict or influence the future use of land so controlled for the benefit of other land in their ownership. *'Land may be owned by industrial concerns or public utilities adjacent to their existing premises to permit expansion even if there are no concrete plans to use this land'* (p. 78). This is illustrated by a recent survey in Greater Manchester which revealed that only 429 out of the 1425 hectares allocated for industrial development by the planning system were actually on the market (Roger Tym & Partners, 1989). Howes (1989) suggests that owners tend to keep land off the market when they are uncertain about their own future needs or where they are unaware of the likely marketability or value of the land. Roger Tym & Partners identified a distinct lack of up-to-date and comprehensive information about the industrial property market in Greater Manchester. Such information shortages are likely to contribute to owner uncertainty.

(b) Costs of relocation

Many firms are deterred from moving to more suitable premises by industrial inertia or the cost of upheaval. Inertia may be caused by potential disruption to production or by the fear of losing skilled and reliable workers. Often, the high cost of new building in comparison with the low market value of existing specialized premises will militate against relocation. As a result, as Fothergill, Monk and Perry (1987) argue: *'firms remain in buildings and locations that would not have been chosen in the light of their current requirements and contemporary market signals'* (p.87). Furthermore, the built up nature of much of the urban landscape can inhibit the transfer of land from low value uses into higher value ones, as neo-classical theory might require. The cost of redevelopment is simply far too high. Firms wishing to expand into adjacent residential areas are discouraged from doing so by the cost and difficulty of acquiring numerous small plots of land from individual vendors.

(c) Planning controls

Howes (1989) contends that, because of substantial imperfections, land markets operate only at partial equilibrium. Supply and demand do not automatically equate. He suggests that supply is determined primarily through the planning system, while demand depends upon private sector action. This makes it extremely difficult to apply competitive market analysis to the industrial land market. As Ratcliffe (1978) makes clear:

> *In a mixed economy, however, the workings of the price mechanism are conditional and controlled by public policy to such an extent that the forces of the free market and intervention become interdependent and at times indistinguishable. The level and pattern of land values, for instance, is as much an expression of prevailing planning policy as it is of open-market demand for and supply of space* (p. 7).

The widespread nature of these several market imperfections challenges the assumption that land will be available wherever the industrialist chooses to locate. A recent report prepared by the management consultants, Deloittes Hawkins and Sells, for Business in the Community, found that 99% of the 129 local authorities contacted reported a shortage of industrial land and premises

(CSW, 1989). If such a national shortage indeed exists, then the supply of industrial land may well be a critical influence on locational decision making. A survey of the availability of strategic industrial sites to meet the needs of potential inward investors to the north-west of England concluded that the lack of suitable sites and premises was one of the most important factors in locational decisions being lost to the north-west (INWARD, 1988).

1.1.3 Equilibrium in land and property markets

In any discussion of the market for land and property, it is important also to remember that it is unrealistic to conceptualize industrial land and property markets as single entities. Numerous submarkets exist, divided principally according to location, size and tenure. Traditionally, heavy manufacturing industry has owned its own premises, often acquired or developed with mortgage finance, while light industry and warehousing have tended to rent speculative accommodation in the investment market (Henneberry, 1988). Most small firms rely on the investment market since they are usually unable or unwilling to finance the construction of their own premises (Perry, 1986). In the owner-occupied market, factories are poor substitutes for one another in comparison, for instance, with residential property. Industry's demand for property is thus extremely varied, whereas the supply is restricted by the narrowness of each particular submarket.

In inner urban areas in particular, empirical work (MacGregor *et al.*, 1985) has shown that the reasons for land market failure may be even deeper than elsewhere. Land as a product is by no means homogenous, buyers and sellers certainly do not possess perfect knowledge and significant externalities exist. In addition, transactions may be few and prices may adjust only very slowly to eliminate surpluses or deficit. Such markets, where they are also characterized by economic restructuring and the flight of capital, have been described as *'fragile'* or *'thin'* (Healey, 1991b).

From a theoretical perspective, these factors therefore call into question the very existence of an 'urban land market' in the strict sense implied by economic theory. Henceforth, the term will only be used in a colloquial sense. From a more practical point of view, where land and property markets fail to function in the smooth and efficient way conceptualized in theory, there will be an opportunity cost in terms of higher unit costs for the individual firm and lower output, employment and international competitiveness for the economy as a whole. For instance, a survey of small industrial and warehousing companies looking for new premises in inner

Manchester found that while 47% had been searching for less than six months, a surprising 37% has been searching for between 12 months and two years and 10% had been searching for more than two years. Furthermore, 63% of these companies stated that their failure to find suitable premises had prevented an expansion which would otherwise have occurred in output, employment or both (Adams, 1987).

A new conceptualization is therefore required which views the supply of industrial land and property, not as subservient to the wishes of those companies making locational decisions, but rather as an important constraint to be incorporated within the decision making process. After reviewing how land market imperfections and failure may influence industrial decentralization, the chapter therefore moves on to consider the precise nature of development constraints and the ways in which they are most likely to be overcome.

1.2 ECONOMIC RESTRUCTURING AND INDUSTRIAL DECENTRALIZATION

The extensive literature documenting the fundamental changes which have taken place in manufacturing industry in the UK from the late 1960s pays little attention to the role of land and property markets in providing the physical space for production. Massey and Meegan (1982) for instance, concentrate on production reorganization, which increases productivity and in the absence of similar output growth, leads to employment decline. Such reorganization may take the form of intensification, rationalization or investment and technical change. Indeed structuralist approaches to industrial location would argue that spatial investment decisions can only be properly understood as part of the wider process of capital accumulation.

Between 1980 and 1986, manufacturing productivity in the UK rose by 26.3% but manufacturing output by only 2.8%. Manufacturing employment fell by 18.6%. This decline however was spatially concentrated. In East Manchester for instance, the traditional industrial area of the city, 20 000 jobs were lost between 1971 and 1985 (equivalent to 60% of the 1971 total). Economic decline now replaced slum clearance and highway blight as the main cause of land becoming vacant and the land market appeared unable to cope with the extensive scale of industrial land recycling required to regenerate the area (Adams, Baum and MacGregor, 1987). Similar stories can be told for most traditional manufacturing areas in the UK. Between 1974 and 1982, Coventry lost 53 000 manufacturing jobs, a fall of 45% (Healey and Clark,

1984). Between 1981 and 1984, manufacturing employment declined by 30% in Middlesbrough, but grew by 5% in Swindon (Cooke, 1989). Hall (1988) characterizes the trend away from capital extensive heavy industry to high technology manufacturing and services as an essential component of Britain's 'industrial revolution in reverse'.

One well documented corollary of employment decline in manufacturing and the relative sectoral shift into services has been the decentralization of industry from the cities. The explanations which have been put forward to account for industrial decentralization are many and varied. Deliberate urban renewal policies seeking to reduce industrial concentration or remove nonconforming users have been blamed by some writers. In some cases, as Massey and Meegan (1982) suggest, firms may be glad to leave, if substantial sums can be obtained from the sale or development of an existing urban site. Pressure may then build up to redevelop redundant and unsightly former industrial premises for residential or retail use, as Watson (1987) found in the Black Country. Other explanations for industrial decentralization include high local tax rates, labour conflict and high levels of unionization in inner cities (Scott, 1982). Scott himself argues, however, that inner areas tend to have a competitive advantage for labour intensive industry while peripheral areas have enjoyed a competitive advantage for capital intensive industry. In this way, industrial decentralization can be linked to the processes of production reorganization described by Massey and Meegan (1982).

For a more detailed explanation of how these processes occur, it is necessary to turn to the variety of writers who link industrial decentralization or the urban—rural shift, to the declining capacity of existing urban areas to sustain given employment levels in the face of rising capital-labour ratios (Keeble, 1968; Fothergill and Gudgin, 1982; Keeble, Owens and Thompson, 1983; Fothergill, Monk and Perry, 1987) . Specifically, Fothergill and Gudgin (1982) contend that the relative availability of industrial land in different types of area contributes to the decentralization of industry. Subsequent research by Fothergill, Kitson and Monk (1983a; b) aggregated land availability information obtained from local authorities for different types of urban areas and attempted to link changes in aggregate industrial floorspace to changes in employment in cities, towns and rural areas. This suggested that available industrial land in 1982 stood at 2.0 hectares per thousand employees in the conurbations, but 5.7, 6.5 and 8.7 hectares respectively in large towns, small towns and rural areas.

By means of detailed research involving interviews with 126 managers of plants in the East Midlands, Fothergill, Monk and Perry (1987) sought to sustain the causal link between land supply and employment change. The results demonstrated that many manufacturing firms experience difficulties with sites and premises and that these prove an obstacle to growth and efficiency. Indeed, Fothergill, Monk and Perry (1987) estimated that three quarters of Britain's manufacturing industry operates on sites where at least half the site is already built up and only one third have vacant land adjacent to their site.

Between 1964 and 1985 average employment densities in manufacturing industry in England and Wales fell from 36 to 21 employees per 1000 square metres of floorspace. As Fothergill, Monk and Perry state:

> *It should be stressed that the reduction in employment density is not something that firms actively plan to achieve. It is a by-product of their attempts to maximize the productivity of their workforce in the face of competitive pressures. It does, however, have major implications for the location of employment because it means that over time a factory of a given physical size tends to employ fewer and fewer workers. It also means that in a town where the overall stock of factory floorspace is static (say because of restrictive planning controls) a gradual reduction in manufacturing employment can be expected* (pp. 29-30).

Redevelopment often exacerbates this trend. Fothergill and colleagues estimate that the replacement of a typical 10 000 square metre, multistorey mill by a modern single storey industrial unit with a 40% site coverage will result in a 90% reduction in floorspace and a 95% fall in employment.

While the work in the East Midlands excluded firms with less than 25 employees, a survey of 77 such industrial and warehousing companies in inner Manchester and Salford, undertaken in 1985, revealed that smaller firms too can find themselves constrained by tight urban locations (Adams, 1987). Many of the companies interviewed experienced difficulties with their existing premises and operated in far from ideal conditions. The most serious problems, however, appeared to have been caused by company growth. This was because most companies took only sufficient floorspace for immediate needs and when business expanded found themselves restricted by lack of space and tightly constrained sites. Until alternative premises were found, growth was inhibited.

Clearly the research work of Fothergill and colleagues has served to raise the importance of land as a factor of production and to indicate the existence of a relationship between supply constraints upon land in existing urban areas and industrial decentralization to small towns and rural areas. Manufacturing industry is less flexible in its use of land than service industry and calls for further study. Manufacturing companies are less likely to realize their preferred land management and development strategies and more likely to experience consequential loss in efficiency and competitiveness. Furthermore, if the sheer lack of physical space makes the decline of manufacturing employment in inner cities inevitable, then this will encourage a spiral of economic decline involving falling purchasing power, declining markets for service industry, a lower local tax base and reductions in public services in these areas.

Much as the work of Fothergill and colleagues appears plausible, it has been criticized both empirically and methodologically. Spencer (1987) with reference to the West Midlands comments that:

> *Such emphasis on land and buildings* per se *ignores the weakness of external investment through newer firms and branch plants, loss of markets and lack of diversification in existing firms where management, design, quality of product and marketing seem strong factors unaffected by premises. Land and premises do not explain the corporate restructuring processes of the largest twenty-six firms in the West Midlands, whose impact on the local economy has been particularly significant... Nor do such approaches explain the closure of newly-built factories with modern equipment.* (p. 222).

Sayer (1982), while acknowledging the existence of an urban—rural shift in manufacturing employment, has warned against confusing causation with association in seeking to explain it. For Sayer, manufacturing employment growth in rural areas may be as much due to the weakly organized labour or the environmental attractiveness of such locations to management as to the ready availability of cheap land. Methodological criticism has also been directed towards the validity of shift-share analysis, and the dependence of the 'constrained location school' upon this technique. Nevertheless, Fothergill and Gudgin (1979) have mounted a strong defence of their use of shift-share analysis, arguing that it is a highly robust technique, whose results are unaffected by either the choice of base year or the level of disaggregation.

Perhaps more crucially, the use of aggregate land availability statistics at a regional and subregional level by Fothergill and later colleagues presents only a snapshot of land supply at a particular point in time. Yet, as the exploratory work of Barrett, Stewart and Underwood (1978) demonstrated, the supply of land must be related to the flows of sites through the development process. Most of the advanced conceptual models of the development process stress its dynamic nature (Gore and Nicholson, 1991; Healey, 1991a). Where the flow of sites is interrupted by development constraints, land vacancy is likely to occur (Gore and Nicholson, 1985). Indeed, as numerous studies have shown (Burrows, 1977; MacGregor *et al.*, 1985; Chisholm and Kivell, 1987; Cameron, Monk and Pearce, 1988; Civic Trust, 1988), in the very cities where Fothergill and Gudgin might lead us to expect land supply constraints to be at their most intense, the highest rates of land vacancy are recorded.

If we are to resolve the apparent paradox of extensive land vacancy co-existing with supply constraints in inner urban areas, then we need to look at whether and at what price that supply is actually made available in the urban land market (hereinafter referred to as the 'market availability' of land). Although a particular site may appear to be available in crude supply terms, planning, physical, ownership and valuation constraints may well prevent its market availability (Adams, Baum and MacGregor, 1988). A national survey of vacant land in urban areas of England undertaken in 1990 showed, for example, that over 17% of vacant land was in various stages of dereliction (Shepherd and Abakuks, 1992). Attention therefore needs to focus on the existence of development constraints and the ways in which they are most likely to be overcome.

1.3 DEVELOPMENT CONSTRAINTS AND MARKET AVAILABILITY

Despite conventional theory, land is not automatically supplied wherever industry chooses to locate. When construction is involved, the process of industrial development is highly complex, involving coordination between developers, financiers, builders, professional advisors and the public sector over a series of stages (Perry, 1986). Development constraints impede the flow of sites through the development process. In some cases, difficulties over planning permission may lengthen the process, while in others, an unhelpful site owner may be a greater barrier to new development. Of course, every site is unique and the relative importance of various constraints will vary from one site to another.

The seriousness of particular constraints is often not apparent without detailed investigation. As a result, such constraints may not be fully reflected in aggregate availability statistics. Development constraints may, however, have a critical influence on whether and at what speed, land which initially appears to be available, can in fact be developed. *'To be marketable, a site must be at that stage of availability where it can be quickly brought forward if there is potential interest in it'* (INWARD, 1988, para 72). Each site which achieves development feasibility therefore not only needs to secure a viable development which satisfies market conditions, but must also obtain planning permission, overcome any particular site problems and pass into active development ownership. Whereas planning constraints appear more prevalent at the urban periphery, the combination of critical physical, ownership and valuation constraints may render industrial development problematic in inner urban areas. Each of these constraints needs to be examined in turn to determine the extent to which they restrict the locational choice of manufacturing industry.

1.3.1 Planning constraints

Much controversy has surrounded the impact of the planning process upon the availability of land for industrial development. Righter (1977) argued that planning policies were often responsible for high closure rates among inner city firms. Some firms were extinguished as nonconforming users, others were refused permission to expand or to diversify, while others still were adversely affected by slum clearance programmes. Many local authorities, however, have now adopted much more flexible development control policies which seek to retain and expand employment within built up areas. In some cases, however, greater flexibility in development control may threaten rather than help retain employment. For *'increased 'flexibility' may in practice mean increased uncertainty surrounding planning decisions, so that developers come to believe that plans may be changed or overridden on appeal and are accordingly encouraged to attach 'hope value' to sites and keep them unused, sometimes for many years'* (Civic Trust, 1988, p.28).

There is some evidence to suggest that the introduction of the Town & Country Planning (Use Classes) Order 1987, which amalgamated light industrial, research and development and many office uses into the single B1 Use Class, has also been detrimental to the supply of land for industrial development. One of the implications of the new Use Classes Order, which is discussed in more detail in Chapter 3, is that local plans no longer differentiate between sites for light industrial use and those for office use. Local planning authorities have less control over the development of industrial sites for high technology or office uses. The distribution of such employment space and its development is now left to market forces. Most B1 development in Greater Manchester, for example, on land previously allocated for industrial use has *'proved to be nothing short of pure office developments'* (Roger Tym & Partners, 1989, p.7). It is important, therefore, to give as much attention to the impact of planning flexibility as to that of planning restrictions.

While urban containment has been a particular feature of the post-war planning system in the UK, Healey *et al.*, (1988) suggest that industrial users may have achieved higher than average success in securing the release of open land for development. They suggest that industrial firms already established locally or capable of mobilizing national or local support are among the interests most favoured by the contemporary planning system. Some local authorities have now adopted sophisticated approaches to incorporate the interests of industrial users into land allocation. Local planning authorities in the West Midlands for instance, have together identified four sectors of the industrial land market for which provision needs to be made. The authorities are thus attempting to ensure that two premium sites of subregional importance are always available and that allocations are sufficient to meet four year's demand for the best urban land and three year's demand for good urban land. Such approaches reflect an increased responsiveness by the local planning authorities towards the needs of industry and suggest that equal, if not greater, attention should be given to other development constraints.

1.3.2 Physical constraints

Physical and infrastructural constraints often reflect the consumption of capital by previous land owners or the fiscal crisis of the state which restricts the ability of the public sector to tackle industrial dereliction and inadequate infrastructure. As Gloster and Smith (1989) make clear:

> *Though the causes of dereliction are many and varied, a common theme that emerges across the country is the way in which land has been exploited by traditional industries and left in a neglected state with little regard for its impact on future generations. Both public and private sector land owners are guilty. The coal and steel industries extracted minerals and left mine workings and slag heaps; statutory undertakers have rationalized their operational land but have not brought it back into good condition for*

> *development. Major public and private sector industrial facilities have been closed leaving problems of deep concrete foundations, machine bases, contamination and the tipping of waste material... In many areas where dereliction is widespread, there is a need to provide new infrastructure to make land suitable for development. Highway accessibility is vital, particularly for industry/distribution. Sites with poor access and visibility will not attract private sector investment. Many are often sterilized, severed by operational railways and canals* (pp. 1-2).

Nevertheless, rather than view physical and infrastructural constraints as absolute, it may be more helpful to examine the process through which such constraints can be overcome. As Cameron, Monk and Pearce (1988) argue *'in the end virtually all constraints should be capable of expression in terms of the cost of overcoming them and of course this (for the owner, developer or government) will vary according to the factor and its context'* (p. 26).

1.3.3 Ownership constraints

There are several competing land uses which can achieve higher residual land values than industrial development. Owners of former industrial sites often prefer to seek their disposal for retail, office or residential use rather than make the site available to an alternative industrial user. Where such speculative behaviour proves unsuccessful, a considerable time period might elapse before the site is again marketed for industry. Attention has also been drawn to the reluctance of some landowners to respond to market mechanisms and offer unused land for sale when prices rise. This may be for a variety of reasons, some well justified commercially, others of more dubious value. Chisholm and Kivell (1987) draw attention to the low holding costs of keeping land unused and to the use of such land as loan collateral. The Civic Trust (1988) highlights the apathy of many owners, both in the public and private sectors, which allows urban sites to fall vacant in the first place. Ownership constraints are less susceptible to financial remedy than physical or infrastructural constraints, and may best be addressed through measures to promote the transfer of land from passive to active ownership, including compulsory purchase procedures, public land registers and waste land notices.

1.3.4 Valuation constraints

Many ownership constraints may well be disguised by valuation disputes. There is, however, some merit in seeing these as separate valuation constraints. Earlier research has suggested that acquisitions by local authorities in the open market and the almost total reliance by professional valuers upon the comparative method of valuation, may serve to support an existing level of prices which would otherwise fall in the context of extensive land vacancy (MacGregor *et al.*, 1985). Subsequent work by Watson (1987) in the Black Country and by Roger Tym & Partners (1989) in Greater Manchester have both confirmed that residual values for vacant industrial land within conurbations can often be substantially less than the prices asked by the site owners. Watson considers that

> *the valuation of disused industrial land such as that in the Black Country poses particular problems. In many cases the land-owners continue to attach a higher valuation to their land than that reflected by the market. These historical or book values are often considerably higher than the prices that are achieved in transactions. The work completed so far suggests that different valuation methods are possibly responsible for sustaining such a situation* (p. 8).

Until site owners come to terms with falling values and reduce their asking prices accordingly or the economy picks up enabling values to rise to meet owners' expectations, a valuation constraint can be said to exist. Land within conurbations may therefore be on the market and appear available for industrial development, but is offered at a price which potential users are unable to afford. In some cases, however, those seeking to acquire potential industrial land for owner-occupation (particularly in order to expand an adjoining factory) may be willing to offer higher prices than speculative developers and may consequently raise vendors' expectations beyond the level at which speculative industrial development is viable.

1.3.5 Overcoming development constraints

The various development constraints discussed above may well impede the availability of land for industrial development but they should not be regarded as either static or permanent. In fact, such constraints are dynamic and may even melt away in the face of an improving property market or

Fig. 1.1 Land and property vacancy: recession in the early 1990s

appropriate public sector action. Fothergill, Monk and Perry (1987) have demonstrated a close link between fluctuations in manufacturing output and fluctuations in vacant industrial floorspace. The fortunes of industrial land and property markets are closely dependent upon the fortunes of the economy as a whole. In his study of Wandsworth, Markowski (1978) found that a slump in demand was the most important reason for continued land vacancy. Developers and investors will not commit funds to industrial development, unless prospective rents justify construction. Research in Greater Manchester showed that, irrespective of development constraints, institutions were unwilling to invest in industrial development until rental levels reached at least £4.50 per square foot in comparison with prevailing levels of £4.00 per square foot or less (Roger Tym & Partners, 1989). Supply constraints should not therefore be examined in isolation from demand pressures, for higher levels of demand may well provide the resources and incentive necessary to overcome particular constraints. As Cameron, Monk and Pearce (1988) argue, however, the difficulty in many of the older conurbations is not simply a lack of demand but rather a lack of confidence in demand.

Such considerations help us to move away from an approach which sees local agencies and industrial concerns as merely compliant to the logic of market processes and development constraints. While these processes and constraints may well be a far more important factor in industrial location than previously recognized, some agencies and industrial concerns are clearly more successful in influencing processes and overcoming constraints than others. It is therefore necessary to set the market availability of industrial land within the debate on structure and agency in land and property development (Healey and Barrett, 1990). Such a framework might help explain why poor quality sites are not necessarily left undeveloped while good quality sites are not necessarily developed. Through ingenuity, resources or recourse to legal powers, certain agents are more able than others, to confront and overcome unfavourable structural conditions. A recognition of the impact of development constraints on the market availability of sites, together with a perspective which acknowledges the active role of those involved in the development process, in confronting constraints and changing the structures within which market forces are brought into play, will lead to a more realistic conceptualization of industrial land and property markets. Locational decision making behaviour will then be seen to be neither totally restricted by, nor entirely free from, market processes or development constraints.

1.4 SUMMARY

This chapter has contended that industrial land and property markets do not work in the smooth and efficient way envisaged by early theorists. On the demand side, industrial location decisions are not necessarily made on an optimizing basis. With reference to land supply, it is evident that the behaviour and expectations of landowners, physical and infrastructural characteristics, the working practices of professional valuers and development agents and complex institutional factors in the form of land use and land supply policies, all play their part in influencing the market availability of land. Earlier research has indicated that spatial variations exist in the distribution of development constraints and it has been argued that these play some part in explaining the urban—rural shift in the location of industry. More recently, however, attention has been focussed on the role of the range of agencies involved in the development process and the extent to which they can effectively challenge structure and overcome constraints.

There is a continuing need, however, for detailed empirical research, exploring the several aspects of current debate and helping to inform the industrial land policies of both central and local government. One such study of the market availability of land for industrial development, undertaken in the north-west of England, is described in the next chapter. Subsequent chapters will draw on the findings of that study to discuss in more detail some of the dimensions of the debate relating to industrial land supply briefly highlighted here. Chapter 3 investigates how land use policy in this area responded flexibly to economic change in the 1980s. The demand for industrial land is examined in Chapter 4 and supply analysed in Chapter 5. Chapter 6 identifies nine important types of interest involved in the industrial land market and Chapters 7 and 8 respectively examine how far the behaviour of these various interests is influenced by the planning system and the physical condition of land. From this analysis, Chapter 9 seeks to differentiate between active and passive behaviour in industrial land development. Chapter 10 draws the main conclusions together and sets out eight types of recommendations for action.

2

THE MARKET AVAILABILITY OF INDUSTRIAL LAND: A CASE STUDY APPROACH

One of the best ways to discover whether land markets and development processes meet the needs of manufacturing industry is to examine how the processes of land supply, land exchange and land development interact in a selected case study location. Only through detailed case work, is it possible to pinpoint development constraints, assess their impact and identify the particular actions which need to be taken, if they are to be overcome.

Research of this nature requires the careful inspection of a wide variety of empirical records as well as structured interviews with planners, valuers, developers, landowners, and of course, industrialists. Its distinctive characteristic is its emphasis on the interaction at the local level between land consumers and the various actors in the development process who can collectively be considered as land suppliers. Although criticisms of case study research have been made in the past, it is now a well tested and generally accepted method of working (Healey, 1986). Of course, care needs to be taken in collecting consistent information, in formulating a structured approach to interviews and particularly in the interpretation of material obtained. Problems of confidentiality may arise but, as Healey and Barrett (1990) maintain *'The sensitive investigative researcher can penetrate development processes with considerable success'* (p.99).

This chapter explains how such research was undertaken in the north-west of England in the corridor from Birkenhead in the Wirral to Crewe in Cheshire (referred to hereafter as the *Cheshire—Wirral Corridor*). In particular, this chapter summarizes the characteristics of that corridor which led to its selection and will consider in some detail the sources of existing statistical information on local industrial land markets. The problems inherent in the use of these sources by those interested in an empirical study of land and property markets will be highlighted.

2.1 THE CASE STUDY LOCATION

The choice of a particular location in which to investigate the processes of land supply, land exchange and land development for industry should approximate, as far as possible, to a local land market. The area definition of local land markets, particularly for industrial development, presents clear theoretical difficulties. Whereas locational decisions in the housing and labour markets are generally spatially contained, manufacturing companies may well make such decisions on a national or even international basis. As an extreme example, the land market for a Japanese car manufacturer looking to establish a European plant may well stretch from Sunderland to Sicily. At the other end of the spectrum, research has shown that most small companies, when searching for new accommodation, are unwilling to move any greater distance than ten miles from their existing location (Adams, 1987).

One approach which overcomes this theoretical difficulty is the application of a two-staged site selection process, described by Fothergill, Monk and Perry (1987, p.108). Briefly, in the first stage, firms will have only a loose idea of preferred location (for instance, 'Scotland'). Only when this is defined more narrowly in a second stage of analysis (for instance, 'within twenty miles of Glasgow') do firms begin to search for specific sites and premises. If this is related to some of the earlier work on industrial decentralization (Cameron, 1973), it would suggest that it might be possible, for empirical purposes, to define a local land market over a distance of say 20—30 miles along a motorway or major trunk route corridor from an established urban centre. Since no single industrial land market exists in practice, such a corridor would allow the interaction between separate submarkets to be examined.

One such corridor in the north-west of England is that approximately 20 miles each side of Chester, along the M53 to Birkenhead and Wallasey in the north-west, and along the A51 to Crewe and Nantwich in the south-east. In practice, the research was extended throughout the Metropolitan District of Wirral and the Cheshire districts of Chester, Crewe & Nantwich and (apart from the Mersey Marshes) Ellesmere Port & Neston, within which the M53/A51 corridor fell (Figs 2.1 and 2.2).

One of the problems inherent in the case study approach is, of course, the extent to which it is possible to extrapolate from the findings of such research. What relevance will a study of a local land market area in the north-west of England have for the wider debate. It could be suggested, for example, that

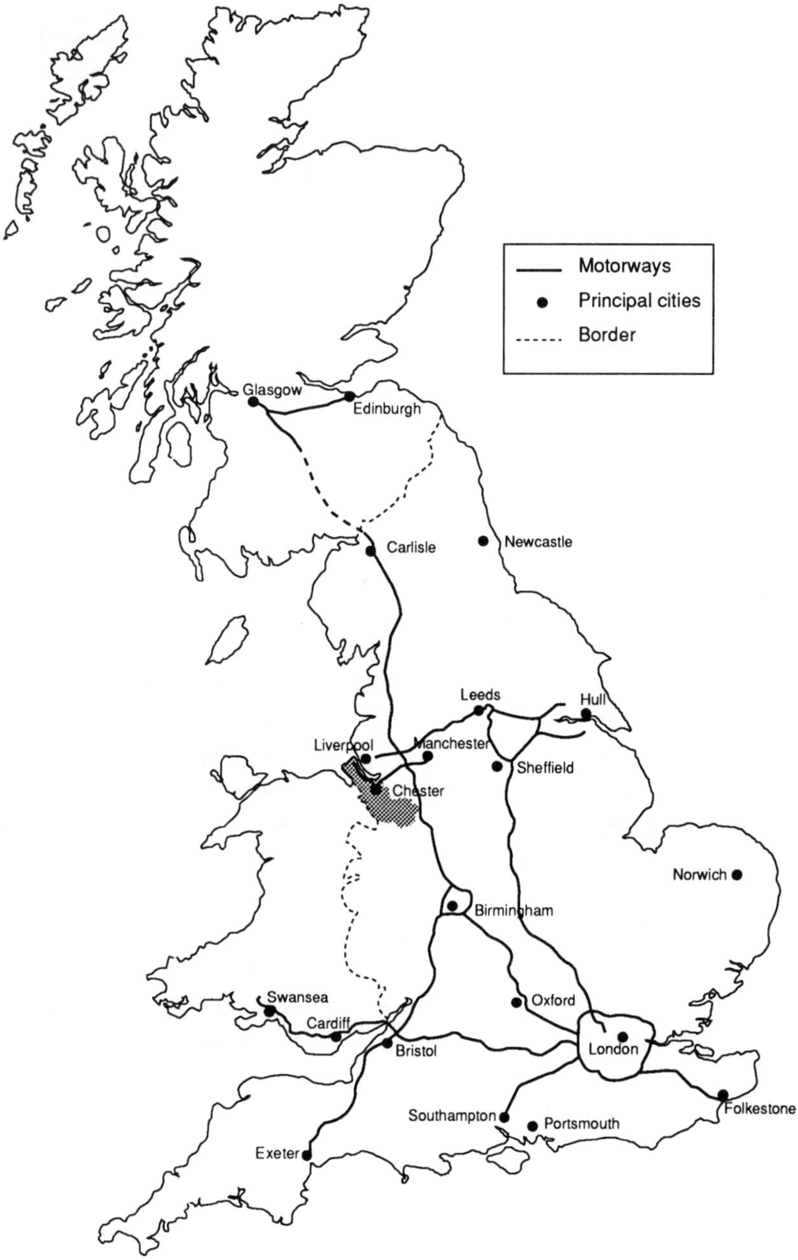

Fig. 2.1 The research location

the north-west region itself is very different from other regions in terms of economic, social and demographic trends; that perhaps we should not ignore the significance of the 'north—south' divide, *'the growth during the 1980s of a distinct gap between the "southern" regions (the South East, East Anglia, South West and in most respects, the East Midlands) on the one hand and the rest of the country on the other...'* (Townroe and Martin, 1992, pp.17-18).

This divide — the lead of the southern regions in terms of employment and output growth, and the resultant differences in development pressure between these two areas — will clearly imply broad spatial differences in local land and property markets.

However, there has been growing awareness that an over emphasis on the 'north—south' divide, or even a regional approach, can be misleading. It provides an imperfect vehicle for the description of patterns of change in the UK over the past 15 years and for the formulation of appropriate national and local policies to address economic recovery or economic inequality. It has been acknowledged that there are important intra-regional differences which call for more subtle and meaningful classifications of spatial areas. Thus, in describing contemporary Britain, Champion and Townsend (1990), for example, have distinguished four types of places — large cities, prosperous subregions, industrial districts and more remote rural areas — as being a more realistic approach than the *'usual regional treatment'*. While regional differences may indeed exist, there is also considerable diversity even within the 'north' and the 'south'.

> *In the south, major differences have emerged between the new buoyant areas of high-tec and service sector expansion, and the previously prosperous but now less dynamic towns that were associated with the manufacturing growth of the 1960s and early 1970s. In the north, while the old industrial towns have suffered the brunt of de-industrialization, the major provincial cities and those towns less formerly dependent on manufacturing appear to have launched on a new service-based wave of growth and regeneration* ... (Townroe and Martin, 1992, p.18).

A more meaningful question, therefore, would be not whether the study area is similar to other local land market areas, but whether the study area in fact gives us sufficient diversity to reflect the wider experience of the rest of the country. Is it thus possible to explore the processes of land exchange and development within contrasting physical, economic and planning environments? Does it exemplify the fundamental changes which have reshaped patterns of economic activity, with their consequent implications for industrial location, and for land and property markets? Does the case study location enable the relationship between inner urban, suburban and rural locations to be considered, for example? As discussed in Chapter 1, earlier research by Fothergill and colleagues sought to establish that the decentralization of manufacturing industry to small towns and rural areas was linked to constraints on the supply of industrial land in established urban areas. In order to explore such questions further, it will be important to be able to consider spatial variations in the operation of planning, physical, ownership and valuation constraints.

It will become evident in this and subsequent chapters, that the case study location does satisfy these research criteria. Its characteristics can be briefly summarized and contrasted under four key headings:

- Physical Environment. The study area includes the inner urban area of Wirral, suburban areas, substantial freestanding settlements such as Crewe and Chester, minor towns and a rural hinterland with smaller villages and a farming community. The pattern and density of established development varies between these areas, and it can be anticipated that the type of physical constraints to new development will also vary. While derelict land and obsolete buildings are characteristic of the inner urban areas and areas of historic industrial development, other adverse ground conditions and a lack of basic infrastructure may inhibit the development of rural greenfield sites.
- Economic and Industrial Structure. The processes of de-industrialization (the decline of employment in the traditional manufacturing sector), tertiarization (the rapid growth of the service sector), and the development of new technology and 'high technology' industries have been central to the changing industrial structure of the British economy in the last 20 years and to the changing pattern of employment in the case study location. Although large numbers of jobs have been lost in recent years in traditional industries such as shipbuilding and engineering in Wirral and Crewe, new industries and the service sector are growing, particularly in Chester and on new peripheral business parks in other locations, attracted by the local environment, communication networks and, in some areas, financial incentives.
- Planning Environment. While economic development objectives have inclined local planning authorities towards policies which facilitate and encourage the development of employment sites and have enabled a measure of rural diversification, for example, the case study location also

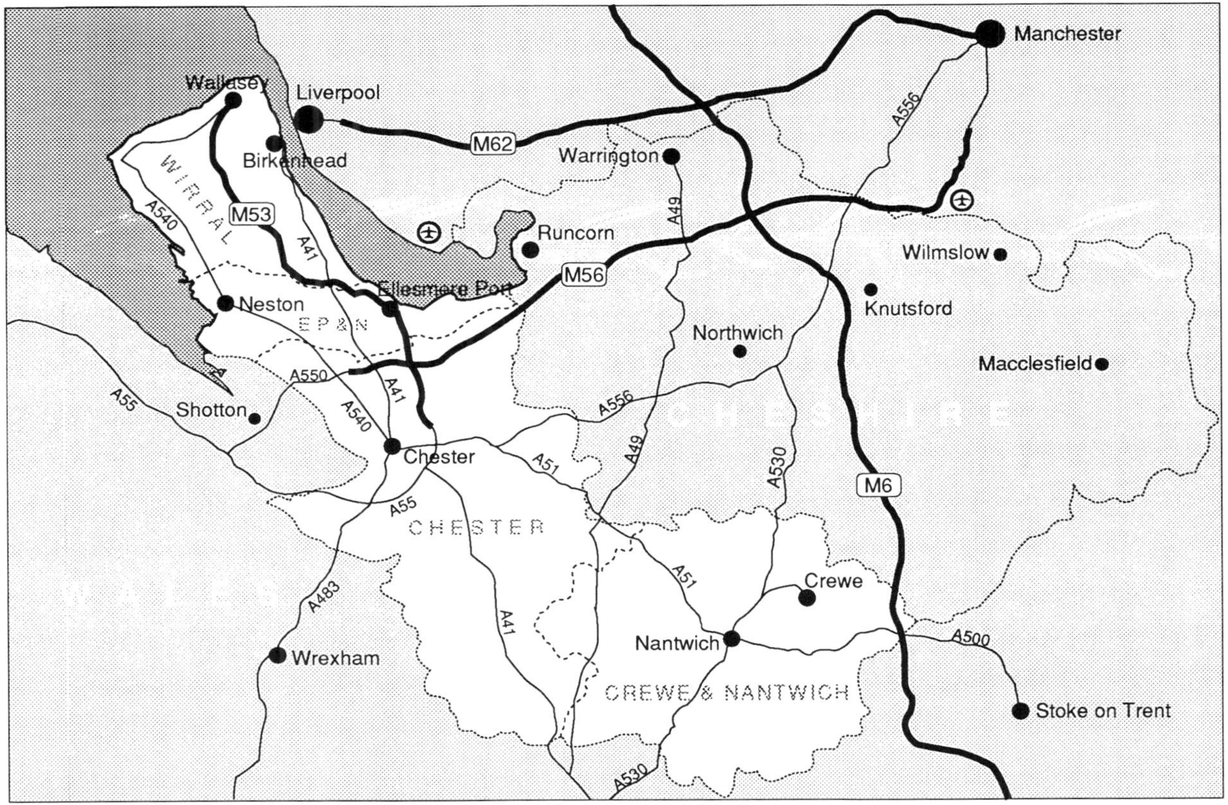

Fig. 2.2 The four administrative districts in the Cheshire—Wirral corridor (note: in Ellesmere Port & Neston, a finger of land to the east, known as Mersey Marshes, was excluded from the analysis of land supply)

includes areas of development constraint defined particularly by green belt policies.

- Economic Development and the Regional and Local Policy Infrastructure. The study location is an area in which a wide variety of policy instruments were at work to influence land supply and industrial location. These included the activities of public development agencies such as English Estates and the Merseyside Development Corporation, part of the latter's designated area being in Wirral, together with a range of financial programmes. The whole of the case study location, excluding Crewe & Nantwich, enjoyed development area status. Regional selective assistance, derelict land grant, urban aid and, latterly city grant, have variously been available in parts of the study area. In addition the separate local authorities have played a key role in promoting economic development.

In summary, the characteristics of this area, and the contrasts within it, make it well suited for a study of the market for industrial land; for an assessment of the impact of supply constraints in different contexts; and for an examination of the role of the various policy mechanisms and agencies which may enable such constraints to be overcome. Later chapters will discuss particular aspects of the study area in more detail, but the following section gives a brief description of the individual districts.

2.2 A PROFILE OF THE CHESHIRE—WIRRAL CORRIDOR

2.2.1 The metropolitan borough of Wirral

This borough covers the northern two thirds of the Wirral Peninsula and is bounded by the River Mersey, the Irish Sea and the River Dee. Birkenhead and Wallasey, the main urban parts of the borough, face across the Mersey towards Liverpool. The original development of what are now the inner areas of Birkenhead and Wallasey can be traced to rapid expansion during the nineteenth century, largely as a result of maritime activity. The decline of both shipbuilding and port related activities, especially in the last twenty years, has resulted in a substantial rise in local unemployment and has been associated with many of the typical inner city problems found elsewhere. The western side of the Wirral Peninsula is more rural, with country parks, farmland and small towns and villages. A good road network serves the Wirral, which includes the M53 and two road tunnels under the Mersey.

Fig. 2.3 The Cheshire—Wirral corridor: an area of contrasts — Wirral docklands

Fig. 2.4 The Cheshire—Wirral corridor: an area of contrasts — Chester

These give access to the major urban areas of Merseyside, Greater Manchester and beyond. In addition, a diesel and electric rail service connects most parts of the Wirral with Liverpool and other parts of the north-west.

2.2.2 The district of Chester

The district of Chester comprises the historic city of Chester and a large rural area supporting village communities intermixed with dairy and arable farming. It is largely a lowland district, bounded on the east by the mid-Cheshire Ridge, which provides excellent walking country, and on the west by the lowlands of Eastern Clwyd in Wales which rise to form the hills of the Clwydian Range. Chester itself dominates the district. Originally a first century Roman fortress, called Deva, Chester soon grew in importance as a centre of communications owing to its location. Today, the approaches to the city from the east, south-east and south are still largely through unspoiled countryside. To the north, north-east and north-west, it is surrounded by industry in the neighbouring districts: motor vehicles, oil, chemicals, steel and aviation. For many years Chester has had the reputation of being a regional shopping centre of distinction. The diversity of shops, especially those under cover in the unique 'Rows', is comparable with much larger cities. As a result, Chester serves not only its own suburbs but also large parts of southern Merseyside, west Cheshire and Clwyd. Chester is one of the country's main tourist centres, with over one and a half million visitors annually, attracted by its Roman history, the Rows, the city walls and the cathedral. As the county town and an important administrative centre, Chester enjoys a varied employment structure and has recently witnessed substantial growth in professional and financial services. It is easily accessible to Manchester Airport (40 minutes by road), and to other parts of the north-west due to its proximity to the M53 and M6 motorways.

2.2.3 The district of Crewe & Nantwich

The district of Crewe & Nantwich lies south-east of Chester and west of Stoke-on-Trent. It comprises the industrial town of Crewe, the historic town of Nantwich and a large rural area supporting farming and village communities. At Crewe, dependence on British Rail and Rolls Royce, the town's two largest employers, has been reduced with the development of newer industrial estates. Nantwich is a locally important market town, with only limited industrial employment. Crewe town centre has been substantially redeveloped, attracting a wide range of shopping and leisure facilities. Nantwich is particularly picturesque, with an Outstanding Conservation Area, containing over 100 listed buildings with many fine Tudor, Georgian and Victorian examples. Tourism has become more important in the district, particularly since the development of the Crewe Heritage Centre, marking the 150th anniversary of the railways coming to the town. Crewe is of course, one of the country's busiest railway junctions with over 400 trains passing through daily. A new motorway link connects Crewe with the M6 motorway, placing the town's main industrial growth area within five minutes driving time of the national motorway network.

2.2.4 The district of Ellesmere Port & Neston

The district of Ellesmere Port & Neston extends across the southern part of the Wirral Peninsula from the Mersey to the Dee. The largest urban area, the town of Ellesmere Port itself, stands beside the Manchester Ship Canal and the River Mersey at the point where these waterways are joined by the Shropshire Union Canal. As in Birkenhead and Wallasey, the earlier development originally took place, the sooner industrial dereliction set in and the greater the current need for reclamation. Today, the local economy of Ellesmere Port is dominated by the Vauxhall car plant (employing 5000) and the Stanlow Oil Refinery (where Shell employ 2000). Recent attention has been given to preserving the local industrial heritage and to promoting the attraction of Ellesmere Port & Neston to visitors and tourists. This is reflected in the National Waterways Museum at Ellesmere Port, which covers all aspects of inland navigation and recreates an original nineteenth century canal port. Although Ellesmere Port itself and the Stanlow oil refinery, immediately to the east, are heavily built up, the western part of the district is largely rural or agricultural. Tourism is important at Neston and is associated with its proximity to the Dee Estuary, where salt marshes are popular with birdwatchers. Ellesmere Port is well served by water and road communications. Motorway links exist to all the major industrial areas of Merseyside, Manchester and the north-west, via the M53 and M56. There are also rail links to the Wirral Peninsula in the west and to Warrington and beyond in the east. All these combine to give good access to both UK and overseas markets.

2.3 THE RESEARCH APPROACH

In order to explore the process of land supply, it was first necessary to compile an industrial land database for the whole of the Cheshire—Wirral corridor. This included all sites of 0.2 hectares or above which had been the subject of a planning application for industrial development between 1985 and 1989. Some of these sites were then developed during this period, others were not. The database also included sites identified by local planning authorities as having potential for industrial development. The database included not only vacant land but also sites with existing buildings suitable for development. In total, the industrial land database covered 872 hectares and 302 different sites, of which 164 hectares (109 sites) were developed between 1985 and 1989 or were under construction at the end of 1989, and 708 hectares (193 sites) remained undeveloped at the end of 1989.

A sample of 150 sites was chosen from the database for detailed investigation. The sample was structured to give an equal number of developed and undeveloped sites and to balance the spatial distribution between four types of area. These were:

- **Inner Urban** The official boundary of the inner urban area in Birkenhead and Wallasey was adopted.
- **Major Urban** This was defined to include the remainder of Birkenhead and Wallasey and other settlements with a population over 35 000 (Chester, Crewe and Ellesmere Port).
- **Minor Urban** Settlements between 10 000 and 35 000 in population were included as minor urban areas.
- **Rural** The remainder of the research location was defined as rural.

One half of the sample was then selected as key sites, from information provided by the local planning authorities, while the other half was selected randomly. The structured nature of the sample thus produced more developed sites, more inner urban sites, more large sites and more sites which had been brought forward by reclamation or other public sector action than would have been the case if it had fully reflected the characteristics of the database. A random sample would have been less likely, however, to address the central issues than one structured in this way — that is the impact of supply constraints on the development process, how they are overcome, their spatial distribution and their relevance to industrial location decisions.

Following the sample selection, the relevant landowner or developer of each of the 150 sample sites (in most cases, a manufacturing company) was approached by letter and invited to participate in the interview programme. A total of 120 interviews were arranged, giving a response rate of 80%. Discussions were also held with local planning officers and property agents about individual sites where relevant. The information sought for each site is listed in Appendix A.

The detailed inspection of planning registers in the four districts, which was a necessary first step in compiling the database, also enabled a separate analysis of all industrial planning activity between 1985 and 1989. This gave some indication of development interest during this period. Recent trends in industrial land and property markets were also examined through the Inland Revenue Valuation Office (IRVO), local property agents and local authority economic development officers. While the focus of this study was the industrial land market, the relationship between this and the industrial property market cannot be ignored. Trends in rental levels will be indicative of the take up of existing space and will suggest the potential for growing or declining development interest both by industrial users and by speculative developers. These components of the research approach therefore provided the context for the detailed investigation of land supply through the sample sites. The problems inherent, however, in the use of these available sources of information in compiling a picture of local markets are discussed in the following section.

2.4 SOURCES OF INFORMATION ON INDUSTRIAL LAND AND PROPERTY MARKETS

No comprehensive source of information exists at a local or national level on the supply and demand for industrial land. A recent report on the Greater Manchester Industry Property Market (Roger Tym & Partners, 1989) identified three important instances, where information flows were said to be deficient or nonexistent. These were:

- Comprehensive market information about property supply and demand, documenting changing levels of vacancy and take up, proposed developments and rental and capital values.
- Well presented and up-to-date information for the conurbation as a whole on available sites and premises. Such information was particularly scarce in relation to new property, high quality property and large units.
- Exchange of information about specific enquiries between different areas

and agencies, especially where the organization first approached was unable to satisfy the enquiry.

These criticisms, although arising in Greater Manchester, could have been made in many parts of the country. Agencies involved in the development of industrial land have constantly to make decisions in the midst of uncertainty. In the Cheshire—Wirral corridor, the agencies principally involved in the industrial development process were manufacturing companies and speculative developers and investors in the private sector and English Estates, Merseyside Development Corporation and of course, the local authorities in the public sector. Only by piecing together information held separately by the local authorities, the district valuers and by knowledgeable regional and local property agents was it possible to construct an overall picture of market activity. Each of these three sources will now be evaluated in turn.

2.4.1 Local authority sources of information

Local authorities have a direct interest in assessing the supply and demand for industrial land and property, both from a planning perspective and as part of their increasing involvement in local economic development. Local authorities maintain extensive records of planning applications and approvals which provide some indication of overall patterns of activity. For instance, the number of industrial planning applications in the Cheshire—Wirral corridor in 1989 was almost double the figure for each of the two previous years. Each of the four district planning authorities kept their own statutory planning register. The information is not recorded consistently, however, between individual local authorities and this can create problems of interpretation when an attempt is made to aggregate the data. Thus when the information from Chester, Crewe & Nantwich and Ellesmere Port was transferred on to the central database kept by Cheshire County Council, a number of inconsistencies inevitably emerged. For example, in some cases where applications were received to extend existing factories, site size was recorded only in relation to the proposed extensions, while in other cases it referred to the entire landholdings within the curtilage of the factory.

Such inconsistencies can often only be resolved by personal inspection of the original application. The ease with which this can be done depends very much on the storage and retrieval systems in individual authorities. At Ellesmere Port & Neston, applications were only retained at the main council offices for two years following determination. Earlier applications had been removed to a depot some two miles away. Special arrangements were needed to visit the depot. At Crewe & Nantwich, applications determined more than two or three years previously were stored in the basement. Advance warning was necessary to retrieve such files. As well as such difficulties at district level, the system used in Cheshire to code planning applications proved incompatible with that used in Wirral. It will obviously be necessary for those interested in this field to make the necessary adjustments themselves.

Planning applications are not necessarily translated into developments. The experience in the study area was that no local authority monitored development completions systematically; rather, efforts were concentrated on the availability of sites at the commencement of the development process and particularly in terms of their known planning status. Cheshire has monitored industrial land availability in the same form since 1982/83 and publishes a 'Land for Jobs' report every June. An Employment Land Resource Database was established in Wirral in 1975 but lapsed in 1980. The first comprehensive survey of employment land in Wirral for ten years was undertaken in 1989. Thus, while some development completions could be identified in Cheshire through inspecting earlier availability reports, this was not possible in Wirral. The shortfall in information on development completions must be overcome either through the sometimes excellent local knowledge of planning officers or often by contacting individual planning applicants, particularly in the case of expansionary development within the curtilage of the site of existing premises. Land for expansion tended to be understated in land availability surveys.

Although it is thus necessary to spend considerable time in cross-checking these various local authority sources and in ascertaining whether planning permissions have been implemented, it is nevertheless possible to derive a reasonable picture of the pattern and extent of land development and take up. This, however, should not be confused with demand. Demand may well remain unsatisfied where supply constraints exist. In Nantwich for instance, very little land was available for the expansion of local firms, until the recent designation of the Nantwich Employment Park. As a result, the town's underlying demand for industrial land had been deflected to nearby Crewe for many years.

One potentially rich indication of demand are the records of economic development enquiries kept by the local authorities. All the local authorities in the corridor maintained such records. These sources provide some indication of overall trends in demand. Considerable practical difficulties, however, prevent records of enquiries from proving reliable indicators of actual demand. There is no standard way of recording such information and there is often considerable inconsistency between one authority and another. Information is often recorded with only limited awareness of various submarkets. It is not always possible to distinguish between enquiries received for industrial, warehousing, retail or office sites. Enquiries often respond to perceived levels of supply or to marketing exercises rather than reflecting changes in real demand. For instance, in 1987 Macclesfield recorded the lowest number of enquiries of any district in Cheshire, despite its evident popularity as an employment location immediately to the south of Manchester. In Cheshire County Council's view, the low number of enquiries recorded in Macclesfield was more a reflection of perceived planning constraints than of low demand. In Chester, there was evidence to suggest that the allocation of 15 hectares of greenfield land for the Chester Orbital Park (now known as Chester Gates) helped stimulate enquiries for distribution space in the locality. In Crewe and Nantwich, a national marketing campaign in 1988 and 1989 led to a substantial increase in enquiries. In contrast, when the borough council ceased American advertising from 1986 onwards, it discovered that enquiries received from the United States tailed off markedly.

Such practical difficulties make it necessary to approach enquiry records with extreme caution. They would also be of greater value if they distinguished between hard enquiries, which include specified and immediate requirements for land or property, and soft enquiries which may merely amount to general expressions of interest in the area. Moreover, some attempt needs to be made to follow up enquiries and identify the proportion which result in an actual sale, letting or development. Indeed, the proportion in any one area may well be small, particularly if several local authorities simultaneously receive the same enquiry from a single company. Despite these reservations, an inspection of the enquiries received in the Cheshire—Wirral corridor helps to reveal some of the characteristics of demand, which will be discussed in more detail in Chapter 4.

In summary, the tremendous variety of information collated by local authorities clearly has the potential to be used much more creatively to enlighten the behaviour of all those concerned with industrial land development. This would require closer cooperation between the various local authority departments concerned, between one local authority and another and between local authorities and other relevant agencies, such as development corporations and property agents. Information would also need to be collected with more rigour and published with greater consistency. This would have a resource cost to local authorities, even if it could be partially offset by contributions or payments received from the private sector. In the end, however, better information is likely to improve the efficiency of local land and property markets and thus underpin the resilience of local economies to wider structural changes.

2.4.2 Regional and local property agents

Many property agents active at the national and international level publish their own specialized market assessments (Scarrett, 1988). Such assessments, however, rarely provide adequate detail at the local level. Agents with immediate local experience often possess the best knowledge about recent transactions, rental levels and yields in their own area. Discussions with nine key regional and local property agents proved extremely valuable in helping to build up an image of the demand for industrial land and premises in the Cheshire—Wirral corridor.

Specialist property agents operate within a close circle: information is readily exchanged between the well established firms. The agents themselves meet fairly regularly both on a formal and informal basis. Often particular agents have moved from one firm to another, as their career has progressed. Specialist property agents can provide the best knowledge of local market conditions and of recent deals. Since market conditions are constantly changing, however, property agents are not well placed to provide historic data or set the local market in a broader perspective.

2.4.3 District valuers

District valuers represent the local arm of the Inland Revenue Valuation Office (IRVO). They undertake a variety of official valuation work, including the assessment of rateable values. In the Cheshire—Wirral corridor, district valuation offices are located in both Crewe and Chester. District valuers have access to a rich source of information since the details of every transaction in land and property in their area must be reported to the local valuation office ('the particulars delivered'). Two important reservations, however, limit its availability for research purposes. First, there may be a delay of up to six months between transactions taking place and the delivery of particulars to the district valuer. Secondly, and much more crucially, individual transaction details remain confidential.

Aggregate statistics are compiled by the district valuer from the particulars delivered and are submitted to the IRVO each April and October. This information is summarized for the UK as a whole and for identified regions, in the IRVO Property Market Report. Inevitably, much useful local information is lost in such a summary. For instance, separate issues of the Property Market Report have included industrial property market values for Birkenhead and Crewe and industrial land values for Chester, but not vice versa. Appropriate official figures for industrial property values in Chester and for industrial land values in Birkenhead and Crewe were only discovered by contacting the local district valuer directly. It may be necessary to seek authorization from London for such information. Staff shortages, at the local level, may also restrict the degree of detail which can be made available.

In the Cheshire—Wirral corridor the district valuers were nevertheless able to provide an indication of trends in industrial land and property values over the whole of the area for the period 1985 to 1989. The reliability of such information, however, may well be undermined if few transactions are taking place. For instance, only two or three transactions per year in industrial land were recorded by the district valuer in Crewe between 1985 and 1988. During the same period, both Chester and Wirral each experienced no more than six transactions each year in industrial land. With such low activity, one unusual transaction can easily upset perceived trends. This is, of course, a widespread problem experienced not only by district valuers, but by all those seeking to monitor trends in industrial land and property markets. District valuers, however, appear more willing to evaluate the strengths and weaknesses of the information available to them, than property agents in the private sector.

2.4.4 Official economic statistics

Earlier researchers have demonstrated a close link between changes in output and in industry's investment in buildings (Fothergill, Monk and Perry, 1987). Unfortunately such data is not available at a sufficient level of disaggregation to enable it to be related to the performance of local property markets. Official economic statistics are designed to monitor trends in the national and regional economies. Such information cannot easily be applied to land and property markets, particularly since the government discontinued the Commercial and Industrial Floorspace statistics in 1985. Official economic statistics may however provide useful contextual material, against which the main trends in local land and property markets can be analysed.

2.5 SUMMARY

In summary then, no single source of available information provides an accurate assessment of the supply and demand for industrial land and property. Although local authorities are potentially the most useful source of data, the lack of consistency and rigour in the way it is handled undermines its effectiveness. Development completions are not necessarily apparent from planning registers or land availability reports. Records of economic development enquiries often fail to eliminate double counting or to distinguish between purely speculative interest and definite property requirements.

Guidance on market trends can be obtained from the district valuers and from local property agents, but such trends may fail to reflect the strength of demand, particularly where the market is dominated by an over supply of vacant premises. Industrial land values are difficult to monitor, if few transactions are taking place. The overall pattern of supply and demand can therefore only be identified by drawing together the most relevant information from each of these sources. If formal arrangements existed to bring all this information together in an organized and consistent way, it would be of enormous benefit to the smoother operation of industrial land and property markets and industrial land development processes.

PART TWO

THE OVERALL PATTERN OF SUPPLY AND DEMAND

3

THE RESPONSE OF LAND USE POLICY TO ECONOMIC RESTRUCTURING

The expansion of economic development has been among the most innovative areas of activity for many local authorities since the early 1980s, sparking off a host of literature (Chandler and Lawless, 1985; Mills and Young, 1986; Sellgren, 1991) and finding explicit recognition as a mainstream local authority function in the Local Government and Housing Act 1989. Local authorities were specifically empowered, under Section 33 of this Act, to undertake a wide range of activities to promote the economic development of their areas.

Economic development activities had previously been financed under the powers of Section 137 of the Local Government Act 1972 or, in some cases, under particular local acts. Section 137, however, did not focus specifically on economic development activity alone and gave local authorities discretionary powers to spend only the product of a 2 p rate on expenditure which would benefit the area or some or all of the residents. Lawless (1989) suggests that approximately two thirds of such spending in the inner cities was in fact directed towards economic development by the mid 1980s. It is argued that the extent to which local authorities were engaged in this activity illustrates the tension then existing between central government's desire to limit local intervention and the concern of local government to stimulate local economic development.

However, under the Local Government and Housing Act 1989, not only was it stated that *'the function of every relevant local authority shall include the taking of such steps as they may from time to time consider appropriate for promoting the economic development of their area'* (Section 33), but the level of expenditure which could legally be incurred was substantially increased (Section 36). Local authorities wishing to make use of the new powers are required to produce a local economic development strategy. This document is subject to local consultation and provides a valuable insight into the particular authority's activities and intentions for economic development.

Economic development strategies are multi-faceted and include a wide variety of activities, not all of which depend on direct financial expenditure. Sellgren (1991) provides a useful classification, dividing economic development activities into four broad areas. These categories relate to promotion and business support, planning and property, transport and mobility, and employment and training. This chapter concentrates upon the second of these areas, that of 'planning and property'.

Sellgren identifies 24 different types of action within the broad category of 'planning and property'. These range from the provision of different types of land and buildings to the relaxation of development control policy and providing key worker housing. According to the authorities contacted by Sellgren, one of the most popular types of action within the category was the identification of sites for development, while the most effective strategy was considered to be the development of new units under 200 square metres, including those designed as nursery or starter units.

As Hobbs (1991) has argued, such activities do not take place in a vacuum, but are the result of a political response to wider economic change. Hobbs reveals, with particular reference to Coventry, how planning activities are continually restructured to accommodate changing economic circumstances. Unemployment rates at the local level are close to the heart of many councillors and, over a period of time, have a noticeable impact on local political priorities. This chapter explores how policies specifically for the development of industrial land, as one important component of the planning and property strategies identified by Sellgren, have responded to widespread economic change.

The response by local authorities within the Cheshire—Wirral corridor serves as a vehicle for this discussion. After having outlined the national planning policy context, the chapter will focus in detail on changes in the local economy and on local planning responses to these changes during the 1980s. It will be shown that as local economic conditions deteriorated during the early to mid 1980s, local authorities developed a more sophisticated understanding of the processes of land supply, land exchange and land development. The chapter will demonstrate how local and strategic planning responded to economic development objectives and to a growing awareness of the land requirements of industry and the effect of development constraints. It will consider the impact of these considerations on the quantity of employment land allocated, its quality and location. It will also look at how planning policies responded to the need for economic diversification in rural areas, and will conclude by outlining some of the other property strategies or policy instruments, outside the planning process, which have been useful in facilitating the development of employment land.

3.1 THE NATIONAL POLICY CONTEXT

When the Conservative government was first elected in 1979, it quickly abandoned the attempts of its predecessor to pursue a local industrial strategy and sought instead to minimize or eliminate what it considered to be unnecessary planning constraints on business activity. In 1980, for example, the government emphasized that *'Local planning authorities should be sensitive to the large number of jobs and the large quantity of economic activity which is locked up in planning applications and should settle their priorities accordingly'* (DoE, 1980, para 11). Subsequently, a series of circulars and other forms of government guidance issued during the mid 1980s urged local planning authorities to facilitate industrial development through speed and flexibility in development control and by the allocation of sufficient land in development plans (DoE, 1984b, 1985 and 1986). These themes were brought together in 1988 in Planning Policy Guidance Note 4 (PPG4) (DoE, 1988a) which stated that:

> *Industrial and commercial development is vital to expansion of the economy and of employment. In particular the Government wishes to help small firms to start up and grow. Local planning authorities should respond constructively to informal enquiries from firms both large and small, handle their applications promptly, and take a positive attitude to development control decision affecting them. Development should be prevented or restricted only where this serves a clear planning purpose and the economic effects have been taken into account. Development control must avoid placing unjustifiable obstacles in the way of any development especially if it is for industry, commerce, housing or any other purpose relevant to economic prosperity... When reviewing and framing development plan policies and land allocations local planning authorities should aim to ensure that there is sufficient land available for industry and commerce and that the variety of sites is sufficient to meet the differing practical needs of the user* (paras 1, 2 and 7).

Revised planning policy guidance, issued in 1992, reiterated the main themes of PPG4, but gave emphasis to three areas. Two of these included the ways in which local authorities could help small firms through the planning system and the need for careful consideration to be given to whether proposals for new development might be incompatible with existing industrial and commercial activities. A third concern of PPG4 (Revised) (DoE, 1992d), was also the need for development plans to take account of both the locational demands of business and wider environmental objectives. Although the locational demands of business were therefore to be a key input in the preparation of development plans, the planning system was also seen as playing an important role in integrating both economic and environmental objectives. In particular, development plans were to take account of the transport implications of new development. In identifying employment sites, plans would attempt to maximize the potential use of transport infrastructure other than roads, while sites located away from urban areas were to be reserved for warehousing uses served by a large number of lorries but with low employment densities.

Central government has thus encouraged local planning authorities generally to provide a variety of employment sites to meet differing practical needs. It also sought, however, to encourage economic regeneration in specific designated areas through the streamlined planning regime found in enterprise zones and simplified planning zones (SPZ). At the beginning of the 1980s, government ministers argued that planning controls had contributed to economic decline and urban dereliction. Under the provisions of the Finance Act (1980) and the Local Government, Planning and Land Act (1980) local authorities were empowered to create enterprise zones, subject to approval by the Department of Environment. One feature of such zones was a simplified planning regime, intended to speed up the processing of planning applications and indeed giving deemed planning consent for many developments. Commentators such as Lawless (1989) have concluded that enterprise zones were successful, insofar as new floorspace was built and in that there was net growth in employment, but have considered that financial incentives and public sector development activity were more influential than the more relaxed planning approach.

In 1987, it was announced that there would be no general extension of enterprise zones in England. Instead, local authorities would be encouraged to declare simplified planning zones (SPZ), enabled by the Housing and Planning Act (1986). Within an SPZ local authorities may give advance planning permission for specific types of developments and individual planning applications which comply with stipulated requirements are not necessary. The relevance of this measure has been questioned, given the existing desire of local planning authorities to facilitate employment development. There were no such zones in the Cheshire—Wirral corridor at the time of the study, but it will be evident from this chapter and later discussion of planning constraints (Chapter 7) that the land use requirements of manufacturing companies have been well recognized in development control and in development plans.

In addition to these various measures, some of which focus on the needs of urban areas, central government has also specifically enabled and urged a more flexible approach to employment development in rural areas. In 1988, the government issued guidance intended to promote rural enterprise and development (DoE, 1988b). On subsequent revision and expansion, this guidance stated that:

> *The range of industries that can be successfully located in rural areas is expanding. Many commercial and light industrial activities can be carried on in rural areas without causing unacceptable disturbance. There are attractions to the firms and their staff in a clean and healthy environment, and there are benefits to the local economy and employment. These firms also help bring new life and activity to rural communities and so are generally welcomed and quickly assimilated* (DoE, 1992c, para 2.12).

Some flexibility is evident even within the green belt. In 1988, for example, central government suggested that redundant farm buildings in the green belt could provide suitable accommodation for small firms or tourist activity. *'The re-use of redundant buildings should not be refused unless there are specific and convincing reasons which cannot be overcome by attaching conditions to the planning permission'* (DoE, 1988d, para 16). Furthermore, Circular 12/91 specifically addressed the re-use of redundant hospital sites in the green belt. Ideally, it was considered that such sites and buildings should be re-used for purposes compatible with the green belt. It was acknowledged, however, that where size, layout and form of buildings made them unsuitable for such purposes, it would be necessary to consider whether special circumstances existed which warranted a change of use. It was considered that putting the sites to beneficial use would be preferable to allowing the buildings to remain empty and the site to become derelict.

Despite increasing attention to the importance of employment development and the needs of industry indicated by these various measures, specific guidance on the allocation of land for industrial development has never been as detailed as that for housing development. Even at a subregional level, strategic planning guidance for Merseyside, for example, merely suggested that local authorities should allocate sufficient industrial and commercial land to provide a wide choice of size, type and location of site (DoE, 1988c). A more profound central government influence upon the allocation of land for industrial development was the revision of the Town and Country Planning (Use Classes) Order in 1987. Most office, light industrial and research and development activities were merged into a single use class known as B1. Activities must be capable of location in a residential area without creating nuisance in order to qualify within B1. This is commonly known as the amenity test.

The introduction of the B1 Use Class had far reaching implications for the planning and development of employment sites. In the first instance, planning permission is no longer required to change a light industrial building to offices, provided that the amenity test is passed. The distribution of employment space between these two uses is consequently a matter for the market. Secondly, there is little point in development plans making a distinction between land for light industrial buildings and land for offices. Many plans now simply identify land for employment uses. Thirdly, local planning authorities have much less control over the decentralization of office uses from town and city centres. It is also easier to redevelop urban industrial sites for high technology and office use. Finally, developers of employment land are no longer protected by the certainty of rigid planning. They need to show ingenuity and flexibility to take advantage of the B1 Use Class. The impact of the new Use Classes Order on industrial development in Greater Manchester has already been noted in Chapter 1. There is also evidence in the Cheshire—Wirral corridor that in recent years both the local planning authorities and speculative developers have promoted peripheral business parks on sites which may previously have been considered suitable for traditional industrial estates, while new development within employment parks has been dominated by office and high technology uses rather than more traditional production space.

The several legislative changes noted above with specific relevance to employment land, must now be seen in the context of the recently enhanced status of the development plan. Section 70(2) of the Town and Country Planning Act 1990 required that, when a planning application was determined, local planning authorities should have regard to the development plan, whenever material to the application, alongside other material considerations. Furthermore, the Planning and Compensation Act 1991 significantly enhanced the status of statutory development plans by introducing the new Section 18A into the Town and Country Planning (Scotland) Act 1972 and Section 54A into the Town and Country Planning Act 1990. This stated that '*where, in*

Fig. 3.1 Business and high technology development at Chester West Employment Park

making any determination under the planning Acts, regard is to be had to the development plan, the determination shall be in accordance with the plan unless material considerations indicate otherwise'. Nonmetropolitan district councils are now required to prepare district wide local plans which, together with structure plans (or unitary development plans in metropolitan areas) form the basis of a plan-led planning system. Although the primary role of the market in bringing forward development proposals is acknowledged, planning decisions must not be arbitrary and the development plan provides the criteria and basis for these decisions.

In summary, although some suggest that in practice this recent legislation will make little difference, planning decisions regarding employment development must be undertaken in accordance with a coherent strategy defined by local development plans. Within such plans, local planning authorities have been encouraged to facilitate industrial and other business development by the appropriate allocation of employment land. They have also been urged to be more responsive in their development control functions. With specific reference to the development of sites for manufacturing use, however, there is some evidence to suggest that the impact of the new Use Classes Order has been to limit the effective supply of land for industrial use, at least in the short to medium term.

3.2 THE CHANGING LOCAL ECONOMY

The fundamental restructuring of the British economy which occurred during the 1980s had a profound effect on the Cheshire—Wirral corridor as a result of its dependence on manufacturing employment and particularly on large scale employers. Early signs of the three types of production reorganization (intensification, rationalization and investment and technical change) identified by Massey and Meegan (1982) were clearly evident during the 1970s. In that decade, the two districts most heavily reliant on the manufacturing sector, Wirral and Ellesmere Port & Neston, witnessed reductions in total employment of 13 200 and 10 800 respectively. In contrast, total employment for the same period grew by 2200 in Crewe & Nantwich and by 4700 in Chester. Unemployment rates for Cheshire and Wirral, however, approximately doubled between 1971 and 1979 and doubled again between 1979 and 1981.

Between 1981 and 1987, total employment in the Cheshire—Wirral corridor fell by a further 10 000 jobs. Much of the decline occurred between 1981 and 1984, with signs of some recovery between 1984 and 1987. As elsewhere in the United Kingdom, this recovery was primarily generated by part time employment and by the service sector. In contrast, manufacturing employment was particularly hard hit during the 1980s. Between 1981 and 1987, manufacturing employment declined by 19% in Wirral, by 13% in Ellesmere Port & Neston, by 7% in Crewe & Nantwich. It grew by 11% in Chester, where numerically it is relatively insignificant (Table 3.1).

Table 3.1 Manufacturing employment change 1981—7

	1981	*1987*	*Change 1981—7*
Wirral	25 800	20 900	-19.0%
Ellesmere Port & Neston	17 500	15 200	-13.1%
Crewe & Nantwich	16 200	15 000	-7.4%
Chester	3 500	3 900	+11.4%
TOTAL	63 000	55 000	-12.7%

Source: Department of Employment (figures rounded to nearest 100).

In 1987, manufacturing accounted for 47% of total employment in Ellesmere Port & Neston, 36% in Crewe & Nantwich, 23% in Wirral but only 8% in Chester. In fact, only Chester has experienced intense pressure for

Table 3.2 Unemployment rates (January 1989)

District	*Area*	*Unemployment %*
Wirral	Birkenhead	27.1%
	Bromborough	15.1%
	Heswall	6.8%
	Wallasey	10.3%
	District as a whole	**14.7%**
Ellesmere Port & Neston	Ellesmere Port	12.3%
	Neston	7.5%
	District as a whole	**10.8%**
Chester	Greater Chester	11.1%
	Chester Rural	5.1%
	District as a whole	**9.1%**
Crewe & Nantwich	Crewe Urban	8.8%
	Nantwich	5.2%
	Crewe Rural	4.8%
	District as a whole	**7.7%**
Merseyside		**16.1%**
Cheshire		**8.3%**
North-west		**11.2%**
Great Britain		**8.1%**

Source: Cheshire County Council (figures refer to those out of work and claiming benefit).

employment development in recent years. The pressure, however, has been primarily for office and for warehouse units rather than for traditional manufacturing development. Unemployment rates at the beginning of 1989 demonstrate the disproportionate impact of recent economic change upon the older urban parts of the Cheshire—Wirral corridor (Table 3.2). Unemployment rates in Birkenhead, for instance, were more than five times higher than those for the rural parts of the Chester and Crewe districts.

Economic trends in each of the four districts clearly reflect contrasts in patterns of historic development. The traditional dependence on those industries in which employment has been most susceptible to the processes of economic restructuring of the last two decades has particularly affected certain parts of the Cheshire—Wirral corridor. The decline of shipbuilding in Birkenhead, for example, provides a dramatic example of how over

Fig. 3.2 Wirral's traditional industrial base: Cammell Laird's site

Fig. 3.3 Cheshire's traditional industrial base: Shell Chemicals' plant at Ellesmere Port

reliance on a single employer can adversely affect the economic fortunes of a whole town. In 1966, Cammell Laird employed 16 000 people. Several subsequent changes of ownership (including almost a decade in the public sector), together with the global restructuring of the shipbuilding industry, saw employment fall to 2500 by the end of the 1980s. In 1990, the whole future of the yard was under threat and despite an intensive local campaign to save the remnants of a once great local employer, the shipyard subsequently closed.

One half of all those working in Ellesmere Port & Neston are employed directly in petrochemicals and in vehicle manufacturing. Taking account of indirect employment, the borough council estimates that 90% of all jobs in Ellesmere Port & Neston are dependent upon these two industries. The two largest companies, Vauxhall and Shell, employ 5000 and 2000 respectively. In Crewe & Nantwich, 35% of all jobs in 1981 were provided either by British Rail or by Rolls Royce. Although for many outsiders, British Rail may be synonymous with Crewe, its employment in the borough has fallen in recent years from a high of 20 000 to under 5000. The rationalization of the British Rail estate left approximately 20—5 hectares vacant within close proximity to the town centre.

Rising unemployment in Chester during the 1980s reflected the closure of the nearby Shotton steelworks and associated industries, immediately across the Welsh border in Deeside. The strong service sector in Chester has further expanded, however, in recent years. Marks & Spencer and NWS Bank (a subsidiary of the Royal Bank of Scotland) both took substantial space on the Chester Business Park in the late 1980s in order to house nationwide administrative operations. In the summer of 1993, an American bank locating in Chester brought a further 800 jobs to the area. More limited manufacturing development has also occurred. For example, the Thomson paper group has constructed a very substantial printing hall on the Chester West Employment Park, where it now employs 450 people. Two American printing companies have also moved into the Chester area, adding to manufacturing employment in the town.

According to Cheshire County Council (1985):

> *The recession (of the early 1980s) had a traumatic effect on Cheshire's major manufacturing industries such as vehicles and chemicals. Nevertheless, Cheshire's economy has shown considerable resilience bearing in mind the challenges it has had to face. Technological change, competitive international trading conditions, increased energy costs and increased growth in its*

> *labour force have all resulted in much greater unemployment. Even in this situation job growth has continued in certain sectors* (para 5.6).

While traditional industries such as petrochemicals, vehicle manufacture and engineering have all shed labour, evidence points to the small scale growth within certain parts of Cheshire of sunrise industries such as electronics, fibre optics and biotechnology. Within the Cheshire—Wirral corridor, much of this growth appears to be centred around Chester and presents local planning authorities with the dilemma of whether to release previously protected green belt land to promote local economic regeneration. Parts of the Cheshire—Wirral corridor clearly have both locational and environmental advantages which are attractive to new industrial development. There is easy access by motorway, for example, into much of the corridor. The M53 and M56 motorways connect Wirral, Ellesmere Port and Chester directly with the national motorway network and Manchester Airport. The M6 skirts the borders of Crewe & Nantwich and, with the recent construction of a link road, these two towns now enjoy a direct connection to the motorway.

In addition many inward investors apparently regard Cheshire as the most environmentally attractive county in the north-west. Even so, there are parts of Ellesmere Port, for example, which share many of the depressing characteristics of inner urban areas such as Birkenhead. Meanwhile away from the inner urban area, the western half of Wirral enjoys a quality of environment comparable to much of Cheshire. This, however, does not prevent inward investors associating Wirral more readily with some of the negative headlines achieved by other parts of Merseyside than with the positive ones achieved by much of Cheshire. These perceptions can be very influential.

In order to take full advantage of locational opportunities and to overcome the particular economic problems experienced in parts of the Cheshire—Wirral corridor, determined efforts have been made to strengthen and diversify local economies. Given the widely felt impact of economic restructuring, no local authority in the corridor is relying on the industries of the past to provide the jobs of the future. Instead they are seeking to support and attract newly emerging economic activity and their stance will be shown to be more aggressive and opportunistic than in the past. The way in which economic development objectives were reflected in the reformulation of local land use policy in the 1980s, is discussed later in this chapter. It will be helpful first to give a brief description of the local political and planning context.

3.3 THE LOCAL POLITICAL AND PLANNING FRAMEWORK

Within Cheshire, eight local districts (including Chester, Ellesmere Port & Neston and Crewe & Nantwich) are responsible for local planning and for development control, while Cheshire County Council has responsibility for structure planning and for economic development on a county-wide basis. Until April 1986, the Metropolitan Borough of Wirral, along with the four other districts in Merseyside (Liverpool, Knowsley, St. Helens and Sefton) formed the lower tier of local government below Merseyside County Council. This higher tier was dominated by the Labour Party from 1981 onwards. When Merseyside County Council was abolished on 1st April 1986, each of the five districts, including Wirral, became unitary planning authorities.

During the late 1980s, only Ellesmere Port & Neston Borough Council, with its dominant Labour majority enjoyed political stability (Table 3.3). The Conservatives lost their control of both Wirral and Chester in 1987 and in 1989 were replaced by Labour as the majority party in Wirral. Throughout the period, the Conservatives consistently enjoyed a slender majority over Labour on Crewe & Nantwich Borough Council. However, the balance of power was held first by the alliance between the Liberal Party and the Social Democrats and subsequently by the Liberal Democrats. While Labour remained the largest single party on Cheshire County Council throughout the period, they required the support of the Alliance/Liberal Democrats in order to secure a majority.

Except in Ellesmere Port & Neston, the late 1980s were thus a time of political turbulence in the Cheshire—Wirral corridor. This heightened the political importance of such factors as local unemployment and ensured greater prominence for the promotion of local economic development. Furthermore, the uncertain nature of local politics in Cheshire coupled with the experience of recession in the 1980s, made its politicians reluctant to turn investment away in order to support the regeneration policies in the adjoining conurbations of Merseyside and Greater Manchester. Rather, the reverse was true. As the 1980s progressed, both Cheshire County Council and the three district councils became increasingly determined to attract and retain employment. For example, both central and local government had tended in the past to accord strong support to the Merseyside and north Cheshire green belts which restricted development in the western half of Wirral, in the Chester area and generally north of the A51. More recently, however, Cheshire County Council, Chester City Council and Crewe & Nantwich Borough Council have all attempted to pursue less restrictive policies on new greenfield development in order to encourage economic development.

Table 3.3 Dominant political representation on the local authorities

	1985	*1986*	*1987*	*1988*	*1989*
Wirral Metropolitan Borough Council	Con	Con	(Con)	(Con)	(Lab)
Ellesmere Port & Neston Borough Council	Lab	Lab	Lab	Lab	Lab
Chester City Council	Con	Con	(Con)	(Con)	(Con)
Crewe & Nantwich Borough Council	(Con)	(Con)	(Con)	(Con)	(Con)
Cheshire County Council	(Lab)	(Lab)	(Lab)	(Lab)	(Lab)

Note: In each case, Table 3.3 identifies the party with the largest number of seats. Where no single party enjoyed an overall majority, the largest party is shown in brackets.

As will be discussed in a subsequent section, this initiated a far reaching debate during the recent preparation and consideration of the Cheshire Replacement Structure Plan, which in turn led the Secretary of State for the Environment to uphold more protectionist plans than the local authorities would have wished.

In the years immediately following local government reorganization in 1974, attention in the Cheshire—Wirral corridor focussed primarily on strategic plan making, where the specific allocation of development land has played an important role as a component part of wider economic development strategies (Table 3.4). Structure plans for Cheshire and Merseyside were approved in 1979 and 1980 respectively, and Cheshire County Council has since made particular efforts to review the structure plan and keep it up-to-date. The three district councils in Cheshire also turned their attention to the process of local plan making during the 1980s. Progress, however, has been patchy. By 1990, only Crewe & Nantwich could claim to have recently adopted local plans. There is now, however, a statutory obligation on these districts to prepare a district wide development plan following the implementation of the Town and Country Planning Act (1990), as amended by the Planning and Compensation Act (1991).

In Chester, the city council adopted a district plan in 1983. Although a

Table 3.4 Development plans in Cheshire—Wirral corridor

	Pre-1974	*1979*	*1980*	*1981*	*1982*	*1983*	*1984*	*1985*	*1986*	*1987*	*1988*	*1989*	*1990*	*Under preparation in 1991*
Metropolitan Borough of Wirral	Birkenhead Town Map Wallasey Town Map Bebington Town Map Heswall Town Map Hoylake Town Map		● Merseyside Structure Plan			● Merseyside Green Belt Subject Plan								Wirral Unitary Development Plan
Ellesmere Port & Neston	Ellesmere Port Town Map Neston Town Map † Green Belt Boundaries † Green Belt Subject Plan †								● Mersey Marshes Local Plan					Ellesmere Port & Neston Local Plan
Chester						● Greater Chester District Plan		● Chester Rural Areas Local Plan						Greater Chester Local Plan
Crewe & Nantwich	Draft Nantwich Town Map † Nantwich Town Centre Policy Map †					● Crewe Inner Areas District Plan							● Crewe Local Plan ● Nantwich Local Plan	Borough of Crewe & Nantwich Local Plan
Cheshire		● Cheshire Structure Plan						● Cheshire Structure Plan, First Alteration						Cheshire Replacement Structure Plan (App'd 1992)

† Informal plans which were never statutorily adopted.

The above table shows the date of adoption for plans adopted after 1974. It also shows plans approved prior to 1974 and plans in preparation in 1992.

revised and updated Greater Chester Local Plan was due to be adopted in 1990, a holding direction issued by the Secretary of State for the Environment on 28 March 1990 prevented the city council taking any further steps towards its adoption. Preparation of a new local plan for Ellesmere Port & Neston (excluding the Mersey Marshes) commenced in 1989. A local plan inquiry was held in 1991. Wirral Metropolitan Borough Council showed little interest in statutory local plan preparation, during the lifetime of Merseyside County Council. However, following the abolition of the county council, Wirral MBC has been required to commence work on a new unitary development plan. An Issues and Draft Part One Policies Report appeared in 1990. In due course, when adopted, the unitary development plan will replace the Merseyside Structure Plan, so far as it applies to the Metropolitan Borough of Wirral.

Having outlined the planning framework within the Cheshire—Wirral corridor, the following four sections will consider how local land use planning has contributed to local economic development through its determination of the potential supply of employment sites; how the changing requirements of industry are reflected in policies on the quality and location of land allocated; the extent to which land allocations acknowledge the importance of development constraints; and how local policies have responded to pressure for economic diversification in rural areas.

3.4 THE QUANTITY OF EMPLOYMENT LAND

At the beginning of the 1980s, strategic planning in Cheshire sought to support the regional policy objective of regenerating the older urban areas of Lancashire, Greater Manchester and Merseyside. Unlike the policies of Merseyside County Council, which sought to attract new investment to promote urban regeneration, the approach in Cheshire was relatively constrained and was determined by local job needs. This contrast is evident in the approaches then taken by the Cheshire and Merseyside structure plans. The Merseyside Structure Plan aimed to *'ensure that a shortage of land does not impede industrial development. An adequate supply of sites, in a range of sizes, tenures and locations, must be available for industry likely to settle or grow on Merseyside'* (Merseyside County Council, 1980 para 7.13). In Cheshire, however, a 'needs-based' approach to the allocation of employment land was retained up to and including the first alteration to the county structure plan in 1985. According to the first alteration:

> *Cheshire aims to make enough employment land available to meet its residents' needs for jobs. Over 1000 hectares of land are likely to be needed in the next 12 years from 1979 to 1991. On past trends, about half the land will be re-used derelict or vacant land. Apart from Warrington and Runcorn New Towns which are established as regional economic growth points, development is aimed at providing jobs for the people of Cheshire and not attracting further in-migration, particularly from Merseyside or Greater Manchester* (para 5.8).

As the decade progressed, the practical limitations of such an approach became all too apparent, not least because there is no guarantee that investment turned away from Cheshire would gravitate towards Merseyside or Greater Manchester, rather than to the south-east of England or to other parts of the European Community. More particularly, however, as the impact of economic restructuring spread throughout Cheshire in the 1980s, local planning authorities became reluctant to support a restrictive employment strategy, and sought instead to exploit the locational and environmental advantages of the area. By the end of the 1980s, 'opportunity' had replaced 'need' in strategic planning in Cheshire. The submitted second alterations to the plan (Cheshire County Council, 1990) summarized the later change of emphasis succinctly:

> *Previously the scale of land allocated was related to need for jobs: it is clear that providing land of itself will not create jobs and that the supply of land should be more closely related to the demand from firms, and the opportunities to encourage new investment in the region. For this to happen increasing emphasis is needed on providing a readily available stock of land based on opportunities for new employment development* (Reasons and Explanations for GS2, p.19).

The county council thus abandoned any attempt to relate employment land provision only to local job needs. Rather, the Cheshire Replacement Structure Plan, approved in 1992, proposed that sufficient land be allocated both to allow development rates in the immediate past to be doubled and to take advantage of any special opportunities likely to arise in the future.

This same philosophy can be seen at district level. In the Greater Chester District Plan in 1983, Chester City Council sought to adopt *'an aggressive policy towards the promotion of employment... to encourage new or expanded industrial, office and shopping developments'* (para 11.17). The draft local plan for Crewe, published in 1988, sought to promote higher levels of industrial development in order to reflect Crewe's increasing importance as an employment centre. In Ellesmere Port & Neston, the borough council published a draft local plan in 1989. According to the council, one of the major economic challenges faced by the borough was to take full advantage of the economic opportunities offered by its good communications, development area grant incentives, range of housing, and accessibility to leisure and other facilities, in order to attract new inward investment.

This policy stance was reflected in the quantity of land which planning authorities sought to allocate. The view of Cheshire County Council is, therefore, that the quantity of employment land made available through the planning system needs to be increased markedly to provide choice to industry and attract investment. A comparison of the quantity of employment land allocated earlier in the decade with that allocated in the late 1980s shows that on a yearly basis there has been an increase for all Cheshire districts with the exception of Chester (Table 3.5). However, land allocation figures approved for Chester in the Cheshire Replacement Structure Plan reflect not local planning objectives, but the intervention of the Secretary of State for the Environment, who modified proposed allocations when approving the plan in January 1992. The submitted Second Alteration had proposed not 100 hectares but 210 hectares for Chester, to take advantage of its locational attractiveness. The decision of the Secretary of State, discussed in more detail later (Section 3.6), effectively overturned plans which both the county and district councils had, to promote Chester as a subregional growth point.

In Wirral the principle of urban regeneration has been retained and the policies in the draft unitary development plan sought to allocate sufficient land to provide a diversity of site types, size and location to meet needs within specific submarkets. Although the allocation of land for the period 1986—2001 on a yearly basis is in fact less than that for 1979—86, it is greater than the annual level of completions achieved during the latter period: 120 hectares had been allocated, but development commenced on only 56 hectares of land between 1980 and 1986. The local planning authority argues that:

> *Past completion rates alone, however, are not a reliable indicator of future land needs in Wirral. The unstable conditions that gave rise to them, including a period of deep recession, make it difficult to quantify the level of new development the Borough would*

Table 3.5 Employment land allocations

	Document	*Period*	*Hectares*
Wirral	Merseyside Structure Plan	1979—86	120
	Wirral Issues and Draft Part One Policies	1986—2001	175
Ellesmere Port & Neston	Cheshire Structure Plan First Alteration	1979—91	100
	Cheshire Replacement Structure Plan	1986—2001	140
Chester	Cheshire Structure Plan First Alteration	1979—91	110
	Cheshire Replacement Structure Plan	1986—2001	100
Crewe & Nantwich	Cheshire Structure Plan First Alteration	1979—91	80
	Cheshire Replacement Structure Plan	1986—2001	200

Note: Figures for Ellesmere Port & Neston, Chester and Crewe & Nantwich for the period 1986—2001 are intended to refer to land to be made available and to approximate to the nearest 5 hectares.

> *"normally" support, or the levels future demand would bring and settle at if sustained growth were to occur throughout the plan period* (Wirral MBC, 1990b, p. 52).

Recent employment land allocations by Wirral MBC can thus be seen as reflecting a level of cautious optimism. They exceed earlier take up of sites and anticipate and provide for economic recovery. They also reflect some uncertainty, however, regarding future levels of demand, and an awareness of the dangers of allocating too much land. If any surplus exists, it becomes much easier for owners of sites allocated for employment development to argue at planning appeals for permission to be granted for higher value uses such as housing development. Land speculation may thus result in less, rather than more, land actually being developed for employment use. In Cheshire, where land or buildings allocated for employment purposes are likely to make an important contribution towards implementing the structure plan, a presumption has therefore been introduced against their development for other purposes. An excessive land supply in any one location may also convince property developers and investors that little potential exists for rental growth. Property development finance may then switch away from areas with an excessive supply of land to areas with a much more restricted land supply. This is recognized by Wirral Metropolitan Borough Council which is seeking *'to avoid creating another situation of over supply which would lead to a fall in attainable rent levels and a consequent reduction of incentives for private sector property investment'* (Wirral MBC, 1990b, p. 56).

In summary, the impact of the recession and economic restructuring throughout the Cheshire—Wirral corridor can be seen in local planning objectives. The quantity of land which local planning authorities sought to allocate reflected:

- the need to allocate a sufficient supply of employment land to attract new development and to enable the expansion of existing business;
- the need to diversify the local economic base;
- the need to allocate a sufficient supply to allow for the different needs of different sectors of the market.

In addition, local planning strategies reflected a more aggressive approach to exploiting opportunities for economic growth, rather than simply being reactive to the need for jobs. In response to these objectives, planning authorities began to acknowledge not only the importance of how much employment land to allocate, but also to be aware of the need to identify land of a quality appropriate to the requirements of a wide spectrum of users.

3.5 THE QUALITY OF EMPLOYMENT LAND

As discussed earlier (Section 3.2), economic problems within the Cheshire—Wirral corridor are concentrated in areas of outmoded urban fabric, often previously dominated by traditional industry. At a regional level, planning policy had previously sought to direct new industry to such locations, in order to tackle the worst concentrations of unemployment. At a local level, these policies create a mismatch, however, between the land available to industry and the land on which it might wish to locate in terms of amenity and environmental quality. In addition sites in areas characterized by dereliction, obsolete buildings and outdated infrastructure, often present a multiplicity of development constraints. Increased attention has therefore been given by planning authorities to the appraisal of development constraints on allocated land and to evaluating their impact on the market availability of identified sites.

3.5.1 Amenity and environmental quality

There is growing evidence of the importance of the local environment in influencing location decisions. In the south-east, for example, new service industries have been found to be attracted to 'high amenity' or prestige areas. Economic strength and environmental quality are thus increasingly related (Fielding, 1989). The problems of physical development constraints and environmental quality and location are, of course, often inextricably linked. Nevertheless, the more selective approach of businesses has increasingly been acknowledged by local planning authorities.

Cheshire County Council (1990) acknowledge that:

> *There is a growing need to make efforts to improve the quality of Cheshire's provision (of land) in terms of range, location and availability to meet the changing demands. At present the County suffers from land which offers little choice, and some land is dogged by development constraints and poor locations... The past ten years, has seen a dramatic change in types of employment development and investment demand, with more selective requirements for land. Special attention needs to be focused on accommodating small firms, high technology industries, and the changing needs of service industries... a significant proportion of Cheshire's employment land is unlikely to prove attractive to inward investors* (p. 9).

Similar conclusions have been reached by Wirral Borough Council which recognizes that manufacturing industry is increasingly seeking better quality locations. Such sites according to the council, may account for only a small proportion of land already allocated for employment use in Wirral. The council therefore acknowledged that *'As companies become more discerning about their operating environments, providing land and premises for industry*

Fig. 3.4 Retention of the natural habitat at Crewe Business Park (Case Study 100)

requires a more careful and tailored approach' (Wirral MBC, 1990b, p. 56). This growing attention to the specific concerns of companies when making location decisions has called for an approach which is significantly different from that adopted in the past. Crewe Business Park, for example, which was brought forward in the late 1980s, has been promoted by the district and county councils as Britain's first 'ecology park'. Development here has taken consideration of both the wider desirability of retaining existing hedgerows, ponds and wildlife, and of the way in which these enhance the business environment.

3.5.2 Development constraints

While urban redevelopment land as a whole is known to suffer from a variety of site constraints, that allocated for industrial use often appears to be the most problematic. Concern about the quality of industrial land was expressed in the Merseyside Structure Plan in 1980. Many allocated sites were not immediately available for use because they suffered from deficiencies in essential services or access, from legal impediments or from such obstacles to development as derelict buildings. In 1990, of the 175 hectares of land then allocated for employment uses in the Wirral Unitary Development Plan: Issues and Draft One Policies, only 65 hectares were considered readily available. Many of these sites were already under offer, while others were considered to be unattractive locations in market terms. The other 110 hectares were constrained by such factors as ownership restrictions, ground conditions and access difficulties. The great majority of these sites were in private ownership. Such sites, even if allocated, cannot usually be developed without substantial treatment.

The draft unitary development plan for public consultation, published in 1992, in fact allocated 200 hectares of land for employment use, which included 17 hectares of land held by businesses for expansion. Of the remaining 183 hectares, 160 hectares included those sites with only minor physical constraints and sites where programmes were already in place to alleviate constraints. There were a further 23 hectares allocated as having potential for employment development but which required more extensive treatment not then programmed. It was uncertain as to whether these sites would come forward for development during the plan period but they were allocated specifically to enable treatment and redevelopment to go ahead if required.

In Cheshire, an employment land audit was first prepared by the county council in 1987, in order to assess the quality of all allocated employment sites. The database includes greenfield, vacant, redevelopment or reclamation sites for industry, warehousing, large retail, office and hotel use. It includes sites over 0.2 hectares or 1000 square metres identified in local plans and draft local plans, as well as sites with planning consent. Five specific criteria were used in order to evaluate each site. These were the length of time which the site had remained undeveloped, the level of investment required to allow development to proceed, the development potential and adaptability of the site, the level of commitment from the public sector to, for instance, derelict land reclamation, and finally the level of market demand. On this basis, sites were identified either as attractive, potentially attractive or unattractive. Subsequently, Cheshire County Council refined this approach and now identifies six categories of undeveloped employment land:

- C1 Sites immediately available and on the market.
- C2 Sites having no constraints which are expected to become available within two years.
- C3 Sites with no serious constraints which are expected to become available within five years or which could become available before 2001.
- C4 Sites with serious physical or other constraints which are unlikely to become available before 2001.
- OS1 Owner specific land which is likely to be developed within five years.
- OS2 Owner specific land which could be developed by 2001.

Information on the supply of employment sites in Cheshire at December 1992 (Table 3.6) reveals that over 20% of potential development land would not be available within the following two years. The problem is particularly severe in Ellesmere Port & Neston where, in the six years prior to the publication of these figures, only 28 hectares had been developed — the lowest total of any district in the county. This is in large part attributed to a legacy of dereliction and ownership constraints. Cheshire County Council considers that success in diversifying the local economy of this district from its existing narrow range of industries will depend on providing high quality employment sites. The borough council also recognized that:

> *The quality and attractiveness of employment land is just as important as quantity if the Borough is to attract economic investment... Of the 386 acres proposed in the Local Plan... 87%*

Table 3.6 Employment land supply in hectares, December 1992

District	*C1*	*C2*	*C3*	*C4*	*Total*
Ellesmere Port & Neston	86	5	27	8	126
Chester	53	2	2	2	59
Crewe & Nantwich	60	14	14	0	88
TOTAL	199	21	43	10	273

Source: Cheshire County Council (1993d).
Note: OS1 and OS2 do not contribute to overall employment land as there is no certainty that it will be developed. At mid 1990, however, owner-specific land totalled 350 hectares in the three districts, out of a total potential supply of 762 hectares.

> *has some kind of development constraint and is not immediately available... the Borough had the second highest concentration of unattractive employment land among Cheshire's eight Districts due to development constraints and poor location or market potential* (Ellesmere Port & Neston BC, 1989, para 7.1).

According to INWARD (1988), the quality of employment land is critical to attracting inward investment. Good quality land can usually be developed quickly: 70% of enquiries received by INWARD between 1985 and 1989 expressed a requirement to be operational within nine months. Irrespective of the quantity of allocated employment land, job opportunities may well be lost if its quality is poor. Assessments of employment land quality have therefore become an integral part of the allocation process within the Cheshire—Wirral corridor. Where an undue proportion of allocated land is considered to be of poor quality, three types of action have been taken.

First, poor quality land has been removed from the allocated total. Cheshire County Council (1990), in the submitted second alterations to the Structure Plan, suggested that land previously allocated for employment development which was economically incapable of development between 1986 and 2001, or which was highly constrained, should be reallocated to another use or should be phased for development after 2001 and given an acceptable temporary use. Secondly, action has been proposed to improve the quality of allocated land. In Wirral, for example, the borough council has argued that, since green belt boundaries will remain unaltered, an extensive programme for the remedial treatment of constrained sites is essential to maintain an adequate supply of employment land. Furthermore, '*the local planning authority will seek to ensure that enough land for five years development is available, free from constraint and capable of being developed immediately, at any one time during the plan period'* (Wirral MBC, 1992, para 4.20). Finally, where constrained land appears to deter employment growth, a rapid and effective way to overcome this may well be to allocate further unconstrained and usually greenfield land for employment development. When such allocations are made, however, considerable local controversy may be generated. This is illustrated by the case of the Greater Chester Local Plan, discussed in the next section.

3.6 THE LOCATION OF ALLOCATED LAND

During the early 1980s, the priority accorded to urban concentration and regeneration in the regional strategy for the north-west, and reflected nationally in green belt legislation, was an important influence on the location of allocated industrial land in the Cheshire—Wirral corridor. On Merseyside, the past emphasis in strategic planning on the need for jobs as a determinant of employment land allocations gave support to policies for the regeneration of such older urban areas as Birkenhead and Wallasey, while containing pressure for new greenfield development elsewhere. For instance, the Merseyside Structure Plan (1980) aimed to:

> ... *encourage investment and employment within existing urban and industrial areas. Where possible, vacant and derelict urban land will be used; this will also have the effect of minimising the amount of good quality agricultural land taken for industry... In future, more industrial development must be in or near inner areas and other areas of highest unemployment* (paras 7.14; 7.15).

Urban regeneration has been retained as the underlying principle in Wirral's unitary development plan. The draft for public consultation (1992) requires that proposals for industrial and commercial development should relate to sites within or immediately adjacent to an existing industrial area and the development should contribute towards urban regeneration. The potential limitations of such an approach in terms of attracting inward investment have become apparent, however, and earlier statements acknowledged that:

> ...*there may sometimes be a conflict between the aims of employment promotion and urban regeneration. Sites within inner areas are*

> *not always capable of fulfilling the requirements of all types of firm, especially where there is a shortfall of land readily available for development. The Local Planning Authority will need to allocate enough land to provide a diversity of site types, size and location and this may involve the allocation of sites in other areas of the Borough* (Wirral MBC, 1990a, para 3.17).

While economic development officers would have welcomed the allocation of sites elsewhere in the borough, the unitary development plan continues to operate a policy of restraint outside urban areas as a means to successful urban regeneration, and places importance on measures to promote the recycling of urban land. The potential for new greenfield development is limited by the tightness of green belt boundaries which the unitary development plan maintains. Specific economic development objectives may thus be constrained by wider planning policies and ultimately by political realities. The impact of public opinion on local politicians, for example, can be seen in the outcome of attempts by Wirral MBC in the early 1980s to market a large greenfield site (Case Study 1) on the urban fringe, first for office and research and development use and later for a high technology business park. These proposals met with such fierce local opposition that they were eventually abandoned. Not least amongst the arguments of local people was the apparent availability of vacant urban sites elsewhere.

In contrast with Wirral where urban regeneration remains a high priority, this has become less influential in the Cheshire districts over the past decade. Strategic planning here no longer supports to the same extent the priority attached at a regional level to revitalizing cities such as Liverpool and Manchester.

Throughout the 1980s, the Cheshire Structure Plan generally sought to concentrate employment development within or adjacent to the main settlement areas. However, in order to promote economic diversification and growth in Cheshire during the 1990s, the county council proposed to release a series of substantial greenfield sites for business park development, in the submitted second alterations to the County Structure Plan. A number of locations in the Mersey Belt and M6 corridor were specified as suitable, including Ellesmere Port, Chester and Crewe. This particular policy was subsequently deleted by the Secretary of State and the quantity of employment land to be made available at Chester, for example, was reduced from 210 hectares to 100 hectares.

Nevertheless, it remains the case that, whereas the first alterations to the Cheshire Structure Plan, approved in 1985, expected half of the county's employment needs between 1979 and 1991 to be satisfied by re-used derelict or vacant land, the Replacement Structure Plan, approved in 1992, implicitly allows for a higher proportion of future employment development in the county to take place on greenfield sites at the edge of existing settlements.

Case Study 1
Proposed greenfield development causes controversy

In 1980, the Merseyside Structure Plan advocated the development of campus out of town offices or research and development facilities at Knowsley and at Upton, Wirral. In 1981, a science park was proposed by Wirral Metropolitan Borough Council on 21 hectares of greenfield land which it owned at Upton Meadows. This proposal generated considerable local opposition. Attempts by the local authority to market the site for high technology development in 1982 and 1983 were unsuccessful, since developers were more interested in general industrial use. In 1984, when outline planning permission was renewed, further public debate occurred. A consultation exercise initiated by the local authority culminated in a public meeting at which unanimous opposition to the development of the site was expressed. Wirral MBC then decided to reappraise earlier options and no active attempts were therefore made to market the site in 1985 or 1986. By 1987, the site was included in the public land register as surplus to local authority requirements.

Following enquiries from potential developers in 1988, the suitability of the site for commercial or high technology development was reaffirmed by four separate council committees. A further consultation exercise again identified considerable local opposition, with a substantial petition submitted against development. As a result, the recommendation from the four committees that the site again be marketed for high technology or business park development, with the possibility of a hotel or leisure facility within the development, was rejected by the full council in April 1989. It was subsequently decided that the long term future of the site should be determined by the Wirral Unitary Development Plan and the site has now been allocated for housing and retail use.

If prestige sites are made available for business park development in parts of Cheshire, they may help overcome the widely recognized regional shortage of sites for inward investors. Cheshire may therefore be well placed to take full advantage of such investment opportunities, as and when they arise. Whether employment development on greenfield sites undermines or underpins the economic prospects of nearby towns and conurbations is, of course, a matter for fierce debate. What is clear, however, is the determination of the local planning authorities in Cheshire not to allow their own economic prospects to be put at risk by any further attempt to steer investment from Cheshire towards the inner urban areas of Greater Manchester or Merseyside. Such a policy would, in the view of authorities in Cheshire, be likely to run the risk of losing the investment altogether from the north-west to the south-east, for example, or elsewhere in Europe. Cheshire's determination to allocate employment land where it appears to be wanted by developers and investors reflects the county's own experience of recession in the early 1980s.

At a district level, those local authorities with particular locational and environmental advantages have sought to allocate new employment sites at the urban periphery. An important 7.5 hectare site was allocated in 1990 on the outskirts of Nantwich, for example, and the Greater Chester Local Plan published in draft in 1988 attempted to remove 400 hectares from the green belt to provide housing and employment land on the edge of Chester. Chester City Council's proposals generated fierce controversy which later gained strategic significance during consideration of the Cheshire Replacement Structure Plan. The reasons for this strategy were complex. On the one hand, the city council wished to relieve pressure on the historic centre of Chester and on the attractive environment of many surrounding villages. On the other hand, the council clearly hoped to meet local employment needs by taking advantage of Chester's potential for economic growth arising from its location and, since 1984, from the availability of regional development grants.

Two strategic employment sites were proposed. One comprised approximately 16 hectares at the end of the M53 and the other 41 hectares opposite the Chester Business Park on the southern fringes of the city. Some 498 formal representations were received, most of which were objections to the plan. The Council for the Protection of Rural England, which described the plan as a flagrant breach of national green belt policy, despatched its national planning officer to argue at the local plan inquiry that Chester City Council had seriously over estimated future development land needs. In contrast, a pamphlet published by Aims of Industry (1989) argued that without further land, the city council's success in stimulating economic development in an area of high unemployment and improved business activity would be halted. Support for the conservation lobby came from the

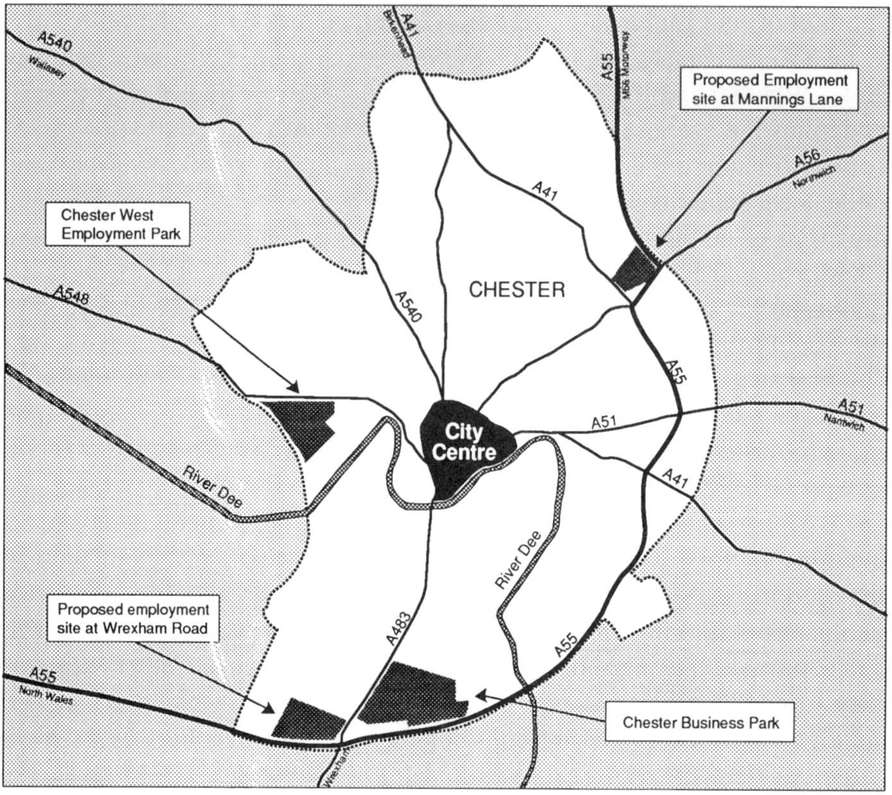

Fig. 3.5 Proposed strategic employment sites within the Greater Chester Local Plan boundary

inquiry inspector who recommended substantial modifications to the plan, including the deletion of the two proposed strategic employment sites. The city council, however, refused to accept these recommendations. In 1990, the Secretary of State for the Environment ordered Chester City Council not to proceed further with the local plan until there had been an opportunity to consider the outcome of the Examination in Public into the Cheshire Replacement Structure Plan.

The Secretary of State's decision on the Cheshire Replacement Structure Plan, and the reduced allocation of employment land in Chester, effectively eliminated these two proposed strategic sites comprising almost 60 hectares in total. Although land use policy has thus attempted to respond to economic change, it has been constrained by wider objectives at national level. Such uncompromising national involvement is unusual. Nevertheless, it indicates that the freedom of local planning authorities to adjust land use policy to local economic circumstances is limited by the extent to which such actions might conflict with national planning guidelines and national political priorities, as well as by local public opinion.

3.7 THE PLANNING RESPONSE TO PRESSURE FOR RURAL DIVERSIFICATION

The response of local planning authorities to pressures for rural economic diversification was a key policy issue in the 1980s. The impetus to promote small scale development in rural areas comes not only from a consideration of levels of unemployment and the number of new jobs which might be required, but also from a recognition of the wider social advantages to rural communities. Almost 45% of the land area of Wirral is within the green belt. Planning policy here, as expressed in the Merseyside Green Belt Subject Plan (Merseyside County Council, 1983), has sought primarily to contain pressure for urban expansion rather than promote rural development. Apart from acknowledging recent changes in national policy on green belts, the Wirral unitary development plan seeks to maintain the earlier policy stance, reflecting the priority still attached to urban regeneration.

In Cheshire, however, much more flexible policies have been pursued since the mid 1980s towards the development of rural enterprise in the districts of Chester and Crewe & Nantwich, whether accommodated in new or refurbished accommodation. In Chester, for example, the Rural Areas Local Plan, adopted in 1985, helped signal this change of emphasis. It was estimated that 1180 people in the plan area would be seeking employment between 1979 and 1991 and, although it was not considered realistic or desirable to provide for this number of jobs in the plan area, it was nevertheless felt to be essential to stimulate the local economy by allocating small sites for light industry and by encouraging existing firms to expand. This plan sought to expand rural employment by designating seven areas within the green belt as employment insets and by allocating a further four rural sites, outside the green belt, for the expansion of existing firms and the establishment of new businesses. A presumption was also introduced in favour of allowing redundant nonresidential buildings to be converted to light industrial use. The southern part of the area covered by the plan is officially recognized as a Rural Development Area, in which grants made available by the Rural Development Commission were instrumental in the construction or refurbishment of small scale industrial premises (Case Studies 2 and 3).

Fig. 3.6 Inset in the green belt: industrial development near Chester (Case Study 2)

Fig. 3.7 Rural diversification in the district of Crewe & Nantwich (Case Study 3)

Case Studies 2 and 3
Economic diversification in the Rural Development Area

These two case studies illustrate the several ways in which planning flexibility and public sector intervention can promote new development in rural areas. Case Study 2, which is in the Chester area, refers to a site allocated in the 1985 Chester Rural Areas Local Plan. Its successful development was therefore first made possible by planning flexibility. Site 3 is within the district of Crewe & Nantwich, but the pattern of its development was parallel with that of Site 2.

These sites were first identified by a Rural Development Area Working Party in 1984. Acquisition was not completed, however, until 1986. Negotiations with both landowners were protracted, particularly for the Crewe site where a condition of sale was that one of the proposed units should be a bespoke building to be let to the landowner. Time spent in finalizing designs and costs helped delay acquisition and development by English Estates, which acted as agents for the Rural Development Commission (RDC). There were further problems in proving title on at least part of each site.

Both sites, which were each less than one hectare, were intended to be developed in two phases. The first phases, ten advance workshops on Site 2, and one factory unit and four workshops on Site 3, were completed in 1987 and were reasonably successful. Development of these first phases was funded in full by the RDC. By January 1990, however, the planned second phases had not been constructed, as a result of a change in priorities of the RDC. The Commission would no longer cover the development costs in full as the level of rural unemployment in Cheshire was not felt to be sufficiently high to make it a priority area.

The residual land was not actively marketed by English Estates, since there was only a limited demand for such sites. The availability of the plots was known, however, both to local agents and to the local authorities. In January 1990 there was active interest in both. On Site 2 the vacant land was expected to be acquired on a 125 year lease by two of the tenants from Phase 1 who were to build units for their own use, retaining some land for future expansion thus vacating their existing four units. The vacant land at Site 3 was expected to be sold on a 125 year lease to a local builder. A planning application...

...proposed seven units, providing a total of 1400 square metres of floorspace. Five units were already earmarked for existing tenants of Phase 1. The subsequent phases of these two developments undertaken by the private sector, have thus enabled the expansion of small companies established in the rural development area. The sites were initially brought forward, however, by public sector intervention which overcame obstacles which may have deterred a private developer — particularly difficulties over site acquisition and a lack of basic infrastructure such as mains water, foul drainage and gas.

Much of the new thinking towards rural economic development was brought together in the submitted second alterations to the Cheshire Structure Plan (Cheshire County Council, 1990). These aimed to take account of the likely estimate of 1000 agricultural job losses in rural Cheshire in the 1980s. Although the north Cheshire Green Belt was retained, it was intended to operate a much more flexible approach towards employment development within the green belt. A more sympathetic attitude towards the expansion or development of existing firms was proposed, in order to help boost the rural economy. The conversion of redundant farm buildings for employment or tourist uses was to be allowed and indeed given priority over conversion to other uses. The Secretary of State for the Environment, in approving the Cheshire Replacement Structure Plan in 1992, deleted these more flexible policies, preferring the county council instead to rely upon the flexibility recently introduced into national green belt policy as discussed earlier (Section 3.1). Outside the green belt, however, the Secretary of State agreed with the county council that new employment development should be encouraged within villages provided that the scale and nature of such development was appropriate.

Another notable issue in rural Cheshire concerns the likely disposal of a number of redundant hospitals. The county council considers that the re-use or, in exceptional cases, the redevelopment of such buildings for employment or tourism uses offers particular benefits. This continued one of the successful themes of the Rural Areas Local Plan (Case Study 4). Although rural Cheshire is no more likely to witness the development of major employment sites in the future than in the past, undoubtedly small scale industry is likely to find it much easier to locate and expand in the rural areas in the future.

Case Study 4
Rural diversification: conversion of redundant hospital

The intended closure of this hospital, which is in the green belt and was built just after World War 1, was announced in the early 1980s. The site was designated as an employment area and an inset in the green belt in the Chester Rural Area Local Plan 1985, and was acquired by a registered charity which provides employment for the disabled with hostel accommodation nearby. Their objectives were twofold: to establish a hospice in the main hospital building (for which funds were still being raised in 1990) and to convert the separate wings into industrial units, hopefully thus affording further employment opportunities for the disabled. The layout of the wards was well suited for conversion to industrial and office premises. Their physical condition was reasonable and it was estimated that no more than £1 per square foot was spent on renovation and conversion. By 1987, the scheme was complete, work having been undertaken in response to demand, and all 3300 square metres of space had been let. The success of the scheme at a time of weak market demand reflected the low rental levels sought, which were also made possible by the nominal sum paid for the site.

The site, however, had poor access, no gas and no mains water. Indeed a major item of expenditure comprised repairs to the water storage tower and replacement pumping equipment. Derelict land grant was critical in covering 80% of the costs of this work. Another special feature of this renovation was the support which it received from the local authority. A special subcommittee, comprising local authority officers, district councillors, parish councillors, professional advisors and the Rural Development Area officer, offered practical help and advice throughout the development process.

3.8 MEASURES TO PROMOTE THE DEVELOPMENT OF ALLOCATED LAND

The allocation of employment land is no guarantee of its actual development. Considerable sums often need to be invested in services and infrastructure and in many cases other development constraints will also need to be overcome. The private sector will only undertake such investment if justified by market demand. Demand, however, is difficult to estimate in advance. In boom times the supply of serviced employment land can quickly dry up. Considerable commitment is often required over a lengthy period of time before the supply is replenished. If demand for serviced employment land cannot be satisfied with the minimum of delay, then the investment opportunities which arise in prosperous times may well be lost from that particular town, city or region. The private sector, however, is rarely interested in speculative investment in infrastructure, in the absence of evidence of immediate market demand. Financial assistance and the direct intervention of the public sector — by both local authorities and development agencies such as the development corporations and English Estates — are two instruments which can address supply side rigidities, helping to overcome development constraints, imperfect knowledge and a lack of confidence by the investment market.

The urban and regional development policies pursued by successive governments have sought to promote the regeneration of Merseyside, including Birkenhead, Wallasey, Bromborough and Ellesmere Port. For many years, these locations have been designated as development areas eligible for regional assistance. In 1984, most of the Chester district was also accorded development area status, in recognition of the high unemployment rates in the Chester Travel to Work Area, as a result of the Shotton closures. The main financial assistance available from central government to the private sector in development areas is in the form of regional enterprise grants or regional selective assistance. Companies in the Chester district, for example, received approximately £1.3 million in regional grants and assistance in the financial year 1987–8 and £3.8 million in the financial year 1988–9. The inner parts of Birkenhead and Wallasey enjoy inner urban area status and were eligible to receive assistance under the urban programme, which was 75% funded by central government (existing funding commitments under the urban programme end in 1994). Private sector development projects in these areas have been eligible for city grant since 1988, when it replaced the earlier urban development and urban regeneration grants. The Merseyside Task Force covers both Wirral and Ellesmere Port, and provides direct central government encouragement to regeneration projects. A separate task force has been established by the Department of Trade and Industry to promote training within the inner part of Wirral, selected for special 'City Challenge' funding.

Throughout the whole of the Cheshire—Wirral corridor, derelict land grant is available for approved schemes on a 100% basis to local authorities and (excluding the inner parts of Birkenhead and Wallasey, where it has been

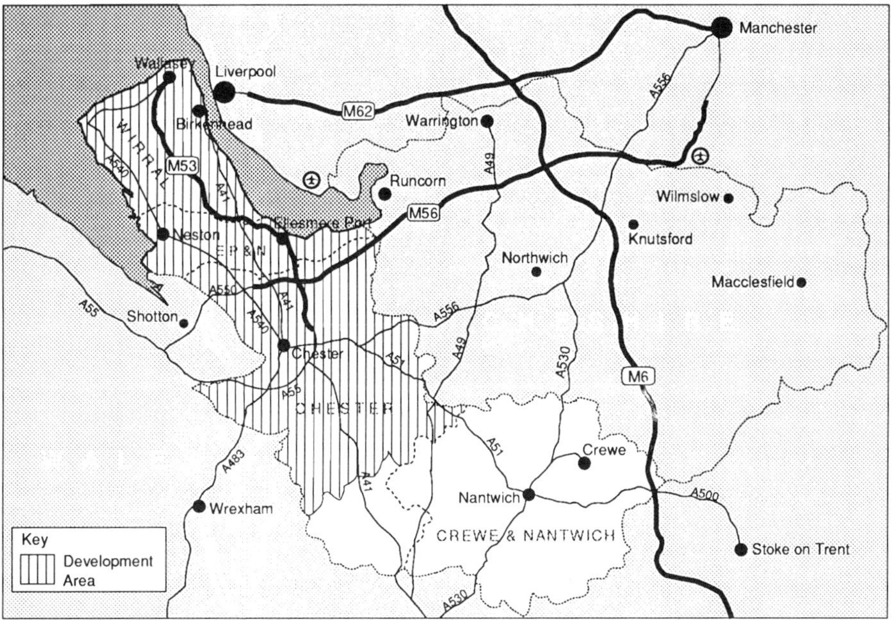

Fig 3.8 Development areas in the Cheshire—Wirral corridor during the study period

incorporated within city grant) on an 80% basis to the private sector. European Community funding is available in the development areas in the form both of low interest loans for the public and private sectors from the European Investment Bank and infrastructure grants to local authorities from the European Regional Development Fund (ERDF). ERDF funding in Wirral now comes through the Merseyside Integrated Development Operation. A wide variety of other minor grants are also available within parts of the Cheshire—Wirral corridor, most notably, those intended to promote rural diversification in the South Cheshire Rural Development Area. The important part played by regional assistance, derelict land grant, the urban programme (as was) and European funding in enabling industrial development will be evident in many of the case studies to be discussed.

Throughout the Cheshire—Wirral corridor, all the local authorities, as part of their economic development functions in the 'planning and property' category defined by Sellgren (1991), have played an active role, either alone or in partnership with the private sector, in ensuring that allocated land is actually developed. Such action has involved both the provision of services and infrastructure on employment sites, and the actual construction of industrial units. Wirral MBC, for example, has maintained a small factory building programme, concentrated in the inner urban areas where it has drawn upon the urban programme and other sources of public finance. Resources, in the public sector, however, became increasingly tight in the late 1980s and new approaches were required to finance the programme (Case Studies 5 and 6). Partnership with the private sector will become increasingly important in response to public sector financial constraints. Partnership is felt by the local authority to be not only desirable but necessary, if the objectives of the unitary development plan with reference to the supply of industrial units are to be achieved.

Case Studies 5 and 6
Financial constraints affect local authority development

These two sites, each of 0.25 hectares, were located within an industrial improvement area in Seacombe, Wirral. The land was previously acquired by the local authority when unfit housing was cleared. Both sites were considered suitable for industrial development, and indeed the local authority proposed to construct advance units itself. No physical constraints existed to prevent development.

Wirral Metropolitan Borough Council needed to obtain financial support and demonstrate that the proposals represented good value for money, before construction could commence. Until 1988, evidence of market demand in this location was meagre. Once a shortfall in supply became apparent in 1989, proposals were brought forward for the development of two 500 square metres advance units on Site 5. However, the cost of the development, at £250 000 could not be met from the local authority mainstream capital programme and the project only proceeded when funding was approved from the urban programme phased over two financial years.

In 1990 the local authority obtained planning permission for ten 100 square metre units on Site 6. Once again, internal budgetary... constraints meant that the local authority had to look beyond its mainstream expenditure to fund the development. An offer submitted for Site 6 by a private developer, who had obtained a separate planning permission, was refused as too low, since at the time land values in inner Birkenhead were felt to be rising rapidly. In fact, budgetary constraints made the local authority consider entering into a joint venture with a private sector partner in order to promote the development of Site 6.

Similarly Ellesmere Port & Neston Borough Council owned 150 industrial premises, mostly in the form of nursery units and small factories, and continued to develop an industrial estate in Neston (Case Studies 7 and 8). The council also acquired and serviced approximately 57 hectares of industrial land between 1976 and 1986.The local authority has made particular use of funds from both the Merseyside Task Force and the European Community to promote the reclamation of derelict land and its servicing for employment purposes (Case Study 9).

Case Studies 7 and 8
Expansion of local authority industrial estate

In 1976/77 Ellesmere Port & Neston Borough Council acquired a further seven hectares of land adjacent to its existing industrial estate in Neston. From 1980 onwards, 36 starter units were constructed by the local authority in a phased development on part of this land (Case Study 7).

The remainder of the land (Case Study 8) which had previously been tipped or contaminated, was reclaimed by the local authority in three phases between 1985 and 1989. Internal access roads and services were also provided. The work was financed by derelict land grants in excess of £600 000.

(Of the reclaimed land, English Estates acquired approximately 2 hectares for the development of workshops and factory units and subsequently expressed interest in further acquisition.)

Case Study 9
Major reclamation by local authority

Ellesmere Port & Neston Borough Council has played a key role in the reclamation and recycling of a former oil refinery site comprising 100 hectares. After the refinery was closed, the local authority acquired the site from Burmah Castrol, together with an adjacent 12 hectares from Manchester Ship Canal Company. The local authority wished to influence the redevelopment of the area and was anxious that it should not become derelict.

About 52 hectares of unreclaimed land were sold to Shell as 'expansion land' although no development has since taken place...

Fig. 3.9 Expansion of local authority industrial estate at Neston (Case Study 7)

...By summer 1990, 20 hectares of the remaining land had been reclaimed with the assistance of derelict land grant. The bulk of this area was to be sold to a resin manufacturer, relocating from three other sites. Reclamation of this first phase and provision of the necessary infrastructure took two years to complete. Reclamation of Phase II, a further 12 hectares, was programmed for 1991/1992. This area is closest to Ellesmere Port and is felt to be suitable for general industry, warehousing and distribution. Other substantial plots have been variously identified for unpopular activities such as scrap metal working.

The local authority had spent £1.8 million of derelict land grant up to the middle of 1989 and required grant aid of a further £1.9 million to complete the reclamation and servicing.

In Chester, the city council took the lead in reclaiming and servicing the 34 hectare Chester West Employment Park, of which approximately 11 hectares is actually just across the Welsh border. In Crewe & Nantwich, the borough council has had a long history of promoting the development of industrial estates, and has thereby lessened the dependence of the town upon its two main employers, British Rail and Rolls Royce. Prior to 1982, the local authority was able to gain access to limited urban programme funds to help construct small workshops on former railway sidings, the purchase and reclamation of which had been supported by derelict land grant. Thereafter, in the absence of urban programme or European Community funding, the council has had to rely on its own resources or enter into partnership with others. One recent joint venture has been undertaken in partnership with Cheshire County Council. This has involved the opening up and servicing of some 35 hectares for the Crewe Business Park, part of which has subsequently been developed by AMEC (Case Study 100). This is a good example of site preparation by the public sector before a boom which has been followed by private sector construction in response to market demand.

As highway authority, Cheshire County Council has played a vital role in promoting better access to employment areas. This is recognized as a key priority of its transport policies. On a smaller scale, funds can be made available from the county council's employment access scheme for much needed minor highway improvements in and around the county's employment areas. In Wirral, where the borough council is also the highway authority, much greater emphasis is given to the accessibility of proposed development sites by public transport. The mixed use nature of much of the borough also makes it imperative to give particular attention to the impact of heavy goods vehicles on local environments, when assessing industrial development proposals.

Action taken by the local authorities has been complemented by similar activity by other public sector bodies such as English Estates and Merseyside Development Corporation. English Estates was also involved in the provision of industrial land and property in the development areas, while the Merseyside Development Corporation is actively promoting the development of former dockland areas in Wirral adjacent to the River Mersey.

3.9 SUMMARY

The severe local impact of wider economic change had a profound effect on planning strategies in the Cheshire—Wirral corridor. A more aggressive approach has developed. The language underpinning specific policies is now that of exploiting opportunities rather than merely solving problems. The objective is economic growth rather than just meeting the need for jobs. Local authorities have been forced to focus much more sharply on the role of land use planning as a strategy for promoting economic development. It has been important to ensure an adequate supply of land to meet potential demand in anticipation of the local economy recovering from the impact of both the recession and economic restructuring. It has been increasingly acknowledged, however, that this strategy alone is not sufficient — that in order to attract inward investment and encourage a diversification of the local economic base, it is necessary to give more careful consideration to the quality and location of allocated sites to meet the wide and different needs of the various sectors of the market.

In those Cheshire districts of the corridor which have particular locational and environmental advantages, and which do not experience the same concentration of problems found in the urban areas of Wirral, rural diversification and the release of new greenfield and green belt sites have been important strands of this strategy. In Chester, enthusiasm for such opportunity based planning proved highly controversial and local objectives have been curbed by central government directives. In Wirral, urban regeneration remained the primary objective throughout the 1980s and early 1990s. Planning policies and land allocations therefore targetted those areas with the most severe social and economic problems, while a tight green belt has been maintained. Despite this different approach in Wirral, the goal is the same. To attract new development, increased importance has been attached to evaluating the quality of land supply, to ensuring a supply of industrial sites to attract new development and to providing a range of sites in terms of size, for example, to meet differing needs. Past investment and existing infrastructure within Wirral's built up areas are seen as offering a number of unique opportunities which a policy of urban regeneration will exploit. These include the M53 motorway, existing well located industrial areas and extensive port facilities.

All districts have therefore recognized the need for a more finely tuned approach to land use policy, if land supply is to meet the requirements of production interests. It is important to emphasize that the details of these strategies have evolved at a local level. Central government maintained a stream of policy advice in the 1980s, including guidance notes on industrial and commercial development and on the rural economy. These, however, provided only a general framework and in some cases emerged concurrently with or after local initiatives. Modifications made by the Secretary of State for the Environment in approving the Cheshire Replacement Structure Plan have in fact reduced the ability of the relevant local authorities to attract inward investment. Whether or not this is justified on other grounds is of course another matter, but it is evidence of conflict between local objectives and national policy stances.

In conclusion, the pattern within the four districts confirms the view of Sellgren (1991) that *'the growing involvement of local authorities in economic development has served to maintain and even to intensify existing trends with a strong emphasis upon activities relating to statutory duties for planning'* (p. 359). Many local authorities see the provision of employment land and its promotion for development as two of the most effective contributions which they are able to make towards the creation or retention of employment. Planning and property related activities were the most important aspect of local economic development throughout the Cheshire—Wirral corridor. The new policy responses which evolved, in turn affected development trends in the boom years of the late 1980s.

4

THE DEMAND FOR INDUSTRIAL LAND

Developers and investors have shown remarkably little interest in the industrial market, compared with their enthusiasm for the retail and office markets. In 1987, for instance, only 9.6% of institutional property holdings monitored by the Investment Property Databank (IPD) were accounted for by the industrial sector. This compares with a higher figure of 16.3% in 1980. According to IPD, total returns from industrial property outpaced those for both retail and office property in only two years during the 1980s — in 1988 and 1989. This coincided with a sharp increase in the demand for industrial land and property owing to economic recovery, following several meagre years in the supply of new speculative development. One significant result of the under-representation of developers and investors in the industrial market, other than in boom years, is that end-users play a far greater role in industrial development than in retail or office development.

This chapter will consider in more detail the relative importance of users, developers and investors in the Cheshire—Wirral corridor, but it will first explore the relationship between the industrial land and property markets and will begin by examining trends in industrial rents. These help provide the context for changing levels of demand for industrial land, insofar as increasing rental levels make new speculative development commercially attractive and also make development by users for owner-occupation a more economic long term solution to their changing production requirements. The impact of the economic cycle, and the recovery in the north-west in 1988 and 1989 from the slump of the early 1980s, will be evident, and manufacturing companies' different strategies for meeting changing production requirements will be highlighted. However, the industrial land market cannot be considered in isolation and a number of case studies have shown the impact on industrial land supply of competing demands for office and retail development.

As already noted in Chapter 2, no single source of statistical information on land and property markets provides an entirely satisfactory indicator of actual demand at the local level. Numbers of enquiries recorded either by property agents or by local authorities can be used to indicate only very general trends. Furthermore, where supply constraints exist, evidence of demand revealed by the take up of space will be diluted. On the other hand, where there is an over supply in the secondary industrial market, which diminishes the rental potential of prime property, rental levels and land values may also provide an imperfect indicator of underlying demand.

The picture presented in this chapter, however, is compiled from information from a whole variety of sources, such as district valuers, local authorities and experienced property agents, particularly active in the Cheshire—Wirral corridor. The data describes how land and property markets were affected by cyclical economic change. Case study material will also be used to illustrate the several responses of consumers of industrial land — whether private or public sector developers or manufacturing companies.

4.1 THE INDUSTRIAL PROPERTY MARKET

Between 1985 and 1987 there was a gradual upward trend in prime rental levels for industrial property in the Cheshire—Wirral corridor (Table 4.1). This was followed by a much sharper rise between 1987 and 1990 and by consolidation or even decline in 1991. By 1991, however, lettings throughout the corridor were few and rent free periods (equivalent, in practice, to a discount on the figures shown in Table 4.1) had become commonplace.

The continuous upward movement in prime rents between 1985 and 1990 is misleading. Business confidence in the area only really improved in 1988. This reflected the buoyancy of the national economy, falling unemployment, low rates of inflation and, until late that year, low interest rates. Subsequent rises in interest rates, leading to recession, caused confidence to fall away. The most important factor behind movements in prime industrial rents between 1985 and 1989 was the persistent shortage of good quality, well located accommodation. The shortage was created by the very low level of new construction during most of the 1980s. Recent development of lettable industrial floorspace in the Cheshire—Wirral corridor is primarily attributable to English Estates and to the local authorities rather than to speculative developers and investors. The shortage was only exposed to the market by the boom years of 1988 and 1989. The rapid increase in rental levels in these two years was therefore the result of a lack of accommodation at a time of increasing demand.

The diversity of demand causes particular shortages of accommodation to

Table 4.1 Trends in prime industrial rents 1985—91

	Birkenhead		*Ellesmere Port*		*Chester*		*Crewe*	
	£ per sq.ft	*Index*	*£ per sq.ft*	*Index*	*£ per sq.ft*	*Index*	*£ per sq.ft*	*Index*
1985	1.90	100	1.35	100	2.25	100	2.10	100
1986	2.00	105	1.50	111	2.25	100	2.30	110
1987	2.25	118	1.75	130	2.50	111	2.65	126
1988	2.60	139	2.00	148	3.50	156	2.95	140
1989	3.00	158	2.25	166	4.25	189	3.75	178
1990	3.50	184	2.65	196	4.00	178	4.50	214
1991	3.50	184	2.65	196	3.75	167	4.25	202

arise in certain submarkets. In 1988 and 1989, smaller units were in particularly short supply in each of the four locations. Chester experienced a shortage of starter units, Crewe & Nantwich and Ellesmere Port, a shortage of units under 500 square metres and Birkenhead, a shortage of units under 1000 square metres. The shortage in Birkenhead subsequently spread to units of up to 4000 square metres. At the other end of the spectrum, agents in Chester also reported a shortage of units of 20 000 square metres and upwards.

Submarkets are defined not only by size of site or premises, however, but also by location. Even within the Cheshire—Wirral corridor smaller local submarkets exist. All four districts experienced rental growth for prime industrial property, particularly during the late 1980s, yet the differentials between districts established by patterns of historic development, supply constraints, locational and environmental advantages persisted. Demand is highly localized since firms will not move far from their existing premises. Ties to the workforce can be very important. Indeed, almost two thirds of the existing manufacturing companies who were interviewed considered the skills of their existing workforce to be the most important factor deterring relocation outside their borough. The pressure of demand within these defined submarkets has a disproportionate impact on rental levels, which is exacerbated by the reluctance of many companies to contemplate the delay of six to nine months likely to be involved in the design and construction of new development.

One local agent in Chester suggested that 85% of deals struck were with companies already based locally. Two thirds of enquiries received by local authority economic development officers in Crewe & Nantwich between January 1988 and April 1989 were from local companies seeking to relocate or expand. Most manufacturing companies only take or build sufficient accommodation for their current needs. When the economy expands, existing premises are often unable to cope with increased business. Action needs to be taken quickly. Unless their premises can be extended, firms look to move to the nearest suitable accommodation. They may be compelled to take property decisions not according to a long term strategy but rather to accommodate short term production requirements, in some cases leading to split site operations (Case Study 10).

Case Study 10
Manufacturing process split between six units

A company which manufactures specialized vehicles such as ambulances and fire engines and which was first established in Chester in 1980, relocated to Crewe in 1982. As a young company seeking room to expand, it was attracted to Crewe by the traditional engineering base and the skilled workforce. It first leased one unit on a modern industrial estate and subsequently leased five additional units nearby as business expanded, the most recent comprising 800 square metres on a 15 year lease in 1988. Each unit accommodates a different aspect of the manufacturing process.

The company has therefore responded to its immediate production requirements by acquiring only sufficient property to meet those requirements. Although the company acknowledges the desirability of a single location, its choice was constrained by its preference to rent rather than to construct or buy its own premises.

In sum, in response to economic recovery in 1988 and 1989, rental levels for industrial property increased substantially in all districts in the Cheshire—Wirral corridor. The market was characterized by a shortage of supply as there had been little confidence in speculative development, at least until the strong market of the late 1980s. There were particular shortages in individual submarkets—most notably in the supply of starter and nursery units—while the locational preferences of local companies which comprise the bulk of market demand, perpetuated differentials in rental levels between the four districts.

4.2 THE INDUSTRIAL LAND MARKET

Between 1985 and 1987, industrial land prices experienced a period of stability, followed by sudden and rapid upward movement (Table 4.2), in contrast to the continuous but more gradual rise in prime industrial rents. Two factors account for this contrasting pattern. One is the fact that far fewer transactions take place each year in industrial land in comparison with industrial property. For example, between 1985 and 1988, district valuers recorded only two to three transactions per year in Crewe and no more than six transactions each year in either Chester or Wirral. One unusually high transaction may be sufficient to initiate a dramatic upward trend in average

Table 4.2 Trends in industrial land prices 1985—91

	Birkenhead		*Ellesmere Port*		*Chester*		*Crewe*	
	£ per acre	*Index*	*£ per acre*	*Index*	*£ per acre*	*Index*	*£ per acre*	*Index*
1985	35 000	100	25 000	100	25 000	100	35 000	100
1986	35 000	100	25 000	100	30 000	120	35 000	100
1987	35 000	100	30 000	120	40 000	160	35 000	100
1988	45 000	129	40 000	160	55 000	220	40 000	114
1989	80 000	229	60 000	240	120 000	480	90 000	257
1990	85 000	243	60 000	240	110 000	440	100 000	286
1991	75 000	214	60 000	240	90 000	360	90 000	257

Note: Actual transactions, which may be few in number, occur in a range either side of these prices.

prices. The second factor, and the most important, is the impact of demand transferred from the industrial property market to the industrial land market.

When shortages of existing property become acute, rents rise rapidly. Demand in the industrial property market is transferred to the industrial land market as both users and developers seek development sites. Some users act directly either specifically to meet changing production needs or to avoid rising rents (Case Study 11). At the same time speculative developers begin to form an expectation that the upward trend in rental levels will soon be sufficient to justify new construction, that speculative development will be economically viable. According to a report from Kleinwort Benson Securities, building costs for a basic industrial unit in the north averaged £30 per square foot, all inclusive, at the beginning of 1991. The rent then required to finance construction was £4.20 per square foot. This was slightly higher than the earlier range of £3.50 to £4.00 per square foot which agents in the Cheshire—Wirral corridor had considered necessary to justify speculative development in 1988 and 1989. Those developers who believe that rents for existing property are heading towards the magic figure necessary to make new construction viable, will turn their attention to seeking suitable sites in the industrial land market.

Case Study 11
Rising rental levels encourage custom built development

A rising property market can have a dramatic impact on the strategies of manufacturing companies. In 1989, a manufacturer of hygiene products discovered that a rent review due the following year, on the small industrial unit which it occupied in mid-Wirral, would be likely to result in a 60% increase in the rent payable. The company formed the view that, with rent reviews due thereafter every three years, its property costs would continue to escalate, and decided to construct a purpose-built unit on a nearby estate, rather than continuing to rent.

A site of 0.22 hectares was purchased for £75 000 by the manufacturer, jointly with another company also likely to face a substantial rent increase on its original premises. Each of these two companies developed its own half of the site. The hygiene manufacturer commenced development of the new factory with ancillary offices in December 1989, with completion expected by the following summer...

...The first phase of this particular estate, owned by a property investment company, had been developed in the 1970s, but a second phase of 10 hectares was granted planning permission in the late 1980s. In fact, by September 1989, the property investment company had been able to sell half of this second phase.

The immediate availability of a building plot within close proximity to the company's original premises was of paramount importance, reflecting the dependence of the company on its local workforce. As a result of the move, the company now owns its own factory, alongside which it has room for a 280 square metre extension when necessary.

Industrial land prices remained generally static between 1985 and 1987, apart from the particular case of Chester. In the following two years, prices more than doubled in Birkenhead, Ellesmere Port and Crewe. In Chester, they increased threefold. Of the four locations, Chester has the lowest proportion of employees in the manufacturing sector and the greatest shortage of traditional industrial land. Chester City Council has been keen to diversify the city's local economy, which has historically been dominated by retailing and tourism, and has recently marketed the city as an attractive environment for inward investment. Land prices in Chester rose rapidly, both as a result of induced demand and restricted supply, but subsequently fell back when inward investment declined with the recession.

Although Birkenhead experienced the slowest rate of growth, land prices moved upwards even here in an area of high unemployment and industrial dereliction. The increase was concentrated, however, in 1989, reflecting the strength of the national recovery in the industrial land market (Case Studies 12 and 13).

In addition to the upturn in the economy and its impact on both manufacturers and developers, two further factors influenced industrial land prices in the Cheshire—Wirral corridor. First the research showed that the development of many industrial sites was facilitated by one form of public subsidy or another. This was not necessarily confined to the inner parts of Birkenhead and Wallasey, areas where the Merseyside Development Corporation was active or where urban programme funding was available. Throughout the corridor, derelict land grant made a critical contribution towards site reclamation. Regional assistance could be sought in most locations apart from Crewe, although from 1988 onwards such grants were only provided on a discretionary basis. According to Cheshire County Council, prior to 1988

Fig. 4.1 Prestige development at Chester Business Park

enquiries from potential inward investors were concentrated in Chester, Ellesmere Port and nearby Halton, all of which enjoyed development area status. After 1988, the greatest growth in enquiries occurred in other parts of the county such as mid and south Cheshire.

Case Studies 12 and 13
Recovery in market demand in the late 1980s

In 1978, English Estates acquired land from British Rail in Birkenhead Docks. The former railway lines were removed, a timber store demolished, and the development of four advance factory units and a number of smaller workshops was completed by 1980. Two smaller parts of the site, however, were left undeveloped, one of 1.25 hectares (Site 12) and the other of 0.4 hectares (Site 13). Since 1982, a sales board on the site advertised the availability of both sites, but throughout most of the 1980s weak market demand for industrial land, particularly in the docklands, resulted in little interest.

Following an upturn in the local property market, plans were brought forward by English Estates in September 1990, to develop Site 12...

> ...The development of eight advance units, of between 325 square metres and 700 square metres was programmed to commence in April 1991, subject to board approval. Meanwhile, in 1989, a private company sought to purchase Site 13 in order to construct a trade and distribution unit. Although the company obtained planning permission for the development in November 1989, it eventually decided to acquire the lease of an existing factory unit adjacent to the vacant plot, and purchased Site 13 for future expansion.
>
> The upturn in the market, even in this part of Birkenhead, was reflected in the asking price of £45 000 — 50 000 per acre sought by English Estates in 1989, in comparison with the lower figure of £25 000 — 30 000 which would have been sought in the mid 1980s. Both figures however, in the view of English Estates, reflect the need to resolve minor ground problems on the site through the limited use of piled foundations.

A second factor which contributed to the upward pressure on industrial land prices in 1988 and 1989 was landowners aspirations in the face of increased competition for land from other uses. Residential land values in the Cheshire—Wirral corridor were between two and three times higher than those of industrial land. One potential industrial site of one hectare lost to residential development was located near Crewe. Planning permission for four light industrial or office units was granted in 1987. The land however was subsequently acquired by a local builder, who constructed 20 dwellings. An even sharper difference was noticeable between industrial land values and those for food retailing. Retail sites could command up to eight times the value of industrial land. When the Crewe Draft Local Plan was under consideration in the late 1980s, approximately 20 hectares of potential industrial land was held off the market, while the owners unsuccessfully sought its re-allocation for retail development, thereby reducing the potential supply of land for industrial development (Case Study 14).

While the factors discussed above helped to reinforce the upward pressure on industrial land values, demand was inevitably diluted in the Cheshire—Wirral corridor by direct competition from nearby locations, which are attractive in terms of land availability or financial incentives. Warrington New Town, to the north, enjoys excellent communications, at the junction of the M6 and M62. The New Town Development Corporation was particularly active in making industrial land available and in seeking to attract firms to

Warrington. Much public sector activity has also been directed towards regenerating the site of the former Shotton Steelworks, immediately across the Welsh border from Chester, which closed in 1979—80. Approximately 120 firms, including Toyota, the Japanese car manufacturers, have been attracted to the area which is now known as Deeside. Both the Welsh Development Agency and Clwyd County Council have been particularly active in promoting Deeside as a major new industrial location. Although there was little direct evidence of firms relocating from the Cheshire—Wirral corridor to Deeside, some of the companies interviewed were prepared to do so in the future, if their expansion needs could not be met at or near their existing locations. Economic development officers in the corridor were clearly envious of the better package of grants which Deeside could offer. In addition, land at Deeside was in plentiful supply, available at approximately half the price sought at nearby Chester, and attractive to inward investment.

Case Study 14
Competition from retail uses for industrial land

This 10 hectare greenfield site, on the edge of an established industrial area of Crewe, had been in the ownership of the adjacent manufacturing company for over 30 years. It was originally held as expansion land but became surplus to requirements. Between 1985 and 1988, it was marketed as an industrial site. At that time, however, there was little demand for industrial land, and particularly for a site of this size.

In 1988, the owners decided to market the site for retail use, since 'out of town' retail centres were achieving far higher land values. However, planning permission for nonfood retail was refused. The Draft Crewe Local Plan, which was then under discussion, had identified three other sites as retail centres, and this particular site was not remarketed for industrial use until adoption of the local plan in 1990.

Although the development of this site for industrial use was initially inhibited by the general decline in the economy in the early 1980s, it can be surmized that its subsequent failure to be developed was the result of the owners hoping to achieve a higher value from retail use, at a time when demand for industrial land was also growing. Indeed, an industrial user had expressed interest in the site during this time...

...It is indicative, however, that the value sought by the agents for industrial use in 1990 was £2.5–3.0 million, compared with the £10 million which might have been achieved for retail use.

Case Study 15
Industrial site sold for service use

Fig. 4.2 Industrial site in Wirral now in service use

This site of 0.6 hectares occupies a prominent corner position in an industrial area in Wirral. The site had previously been occupied by a gasometer and subsequently by a builders yard. After the latter use ceased, the vacant site was left with an uneven covering of rubble. The site was subsequently marketed for some time through estate agents without any success.

A sale of the site was eventually negotiated at the end of 1989 but not for industrial development. The prospective purchaser, a major local garage, intended to construct a service workshop on part of the site and store new cars in a secure compound on the remainder. The local planning authority was enthusiastic to approve this development, to resolve problems of congestion on the streets surrounding the company's other site. The sale achieved £95 000 compared with an industrial land valuation of £85 000.

In summary, the impact of economic recovery and of the take up of available industrial space can be seen in increasing industrial land values in the late 1980s. Demand was reinforced by the availability of financial assistance in the area although there was competition from neighbouring locations. There was also upward pressure on land values, however, as a result of the introduction of the B1 Use Class and as a result of competition for land for retail and other higher value uses, which restricted the potential supply of industrial land. The supply of industrial land will be discussed in more detail, however, in the next chapter.

4.3 SOURCES OF DEMAND FOR INDUSTRIAL LAND

Analysis of the planning applications for industrial development submitted in the Cheshire—Wirral corridor between 1985 and 1988 provides a useful indication of the relative importance of users, developers and the public sector in the supply of new and refurbished industrial accommodation (Table 4.3). It reveals that manufacturing companies were far more active in the provision of industrial accommodation than either the public sector or development and construction companies. Each of these various sources of demand will now be considered in turn.

Table 4.3 Industrial planning applications 1985—8

	No.	*%*
Major manufacturing companies	117	33
Minor manufacturing companies	83	24
Public sector	32	9
Development and construction companies	17	5
Other	103	29
TOTAL	352	100

Notes: (1) The data covers all industrial planning applications, including changes of use, refurbishment and extensions (of 1000 square metres and above).

(2) This analysis excludes minor extensions of less than 1000 square metres.

(3) 'Other' includes private individuals seeking planning permission for small craft and other workshops and self-employed business persons.

4.3.1 Manufacturing interests

The behaviour of manufacturing companies in the land and property markets will depend not only on their immediate production requirements, but also on factors such as their size and corporate structure. These will influence their ability to hold land for expansion purposes, the scale of the market areas within which location decisions are made, and the extent to which management strategies also reflect the value of their landholdings in a wider asset portfolio. Indeed it will be shown elsewhere that manufacturing companies may take on multiple functions in the land market — acting not only as production interests but also investors and even speculative developers. A somewhat crude but practical distinction was therefore made between major manufacturing companies and minor manufacturing companies in the Cheshire—Wirral corridor.

Major manufacturing companies were those listed in independent publications as belonging to the top 20 000 companies in the UK. The research showed that major manufacturing companies rarely needed to purchase land in the market, and only then to establish a new regional site or a new subsidiary of the company (Case Study 16). It was certainly true that, in the case of expansionary development, most major manufacturing companies were able to build on land which they had long held for that particular purpose and rarely needed to acquire additional land for new development. Two major companies in Wirral, Unilever and Premier Brands, provide examples of this. During the 1980s, Unilever invested over £100 million in rebuilding and modernizing its industrial premises in Merseyside. Premier Brands, which manufactures Cadbury products under a franchise in north Wirral, invested over £20 million in extending their factory premises in the late 1980s.

Case Study 16
Acquisition of land by major manufacturing company

Unusually, this is an example of a major manufacturing interest locating a new subsidiary in an inner urban area on a site which afforded sufficient space not only for immediate substantial development but also for future expansion.

A site of 1.8 hectares in Wirral was purchased from the Merseyside Development Corporation in 1987 by a new subsidiary of a major...

...American chemical company already located nearby. The company took a 99 year ground lease at £20 000 per annum, with a peppercorn rent being paid in the first twelve months. It constructed 4500 square metres of new industrial floorspace on part of the site, completed by July 1989. This would meet its production requirements for five years, and the company hoped to expand on to the remaining land by the late 1990s.

The whole of the site suffered from poor ground conditions and drainage problems, as it had previously been used for land fill. Piled foundations were necessary at a cost of £70 000. This additional cost was met by the development corporation who also agreed to cover any similar costs on the land retained for later expansion.

Regional assistance of over £1 million was also forthcoming for the development from the Department of Trade and Industry. The availability of financial assistance from the public sector was undoubtedly the most important factor to influence the company's locational decision making process. The role of the development corporation was also critical in reclaiming an inner urban site which had been vacant for many years.

Minor manufacturing companies are important players in the industrial land market, since they rarely hold extensive reserves of land for future expansion. Such firms may well find it cheaper to acquire second hand premises, when property prices are low. A shortage of good quality secondary accommodation forces prices up and encourages minor manufacturing companies to undertake construction themselves. Demand is generally concentrated on sites of 0.8 hectares or less.

Case Study 17
Expanding company constructs new unit

This case study illustrates how the requirements of a small manufacturing company and its strategies have changed as it has become more established.

In 1989 a plot of 0.21 hectares on an industrial estate in Wirral was acquired from the local authority by a local firm of toolmakers. This...

> ...company was established over 20 years previously and had operated since 1975 from premises which had been designed as a warehouse. This had been built in the 1960s, was uneconomic to run and the layout was less than ideal as a production facility. The present plot was acquired on a 99 year lease and construction of a new factory unit comprising 630 square metres was completed the following summer. The company was thus able to expand, to relocate its production activity to a modern, purpose built unit and felt it had also improved its image.

4.3.2 Public sector development interests

In the public sector, both English Estates and the local authorities were active in assembling land for subsequent private sector development and in undertaking the construction of starter and nursery units. This has been particularly important in Wirral and at Ellesmere Port, where the public sector is an essential provider of modern units to let. In both these locations, extensive public sector involvement has been facilitated by European and central government funding. English Estates, which had its origins in the north-east over 50 years ago was a government sponsored agency whose main duty in broad terms was to further the government's regional and industrial policies, particularly through the management and provision of industrial and commercial sites and premises. It was the intention, certainly since 1981, however, that it should not be active in areas where it was satisfied that sufficient private sector activity was being undertaken. It clearly had a key role therefore in the north-west, an area which is not particularly attractive to speculative developers.

In the inner parts of Wirral, the Merseyside Development Corporation has also been active since 1981. The development corporation pursues its objective of improving the industrial and commercial infrastructure of the inner city in a number of ways—by developing new floorspace or refurbishing and converting existing floorspace; by the provision of financial assistance to companies; by site assembly and reclamation; and by making land available for private development. In March 1992, its landholdings in Wirral totalled 85 hectares, and it too was therefore a key player in the local land market.

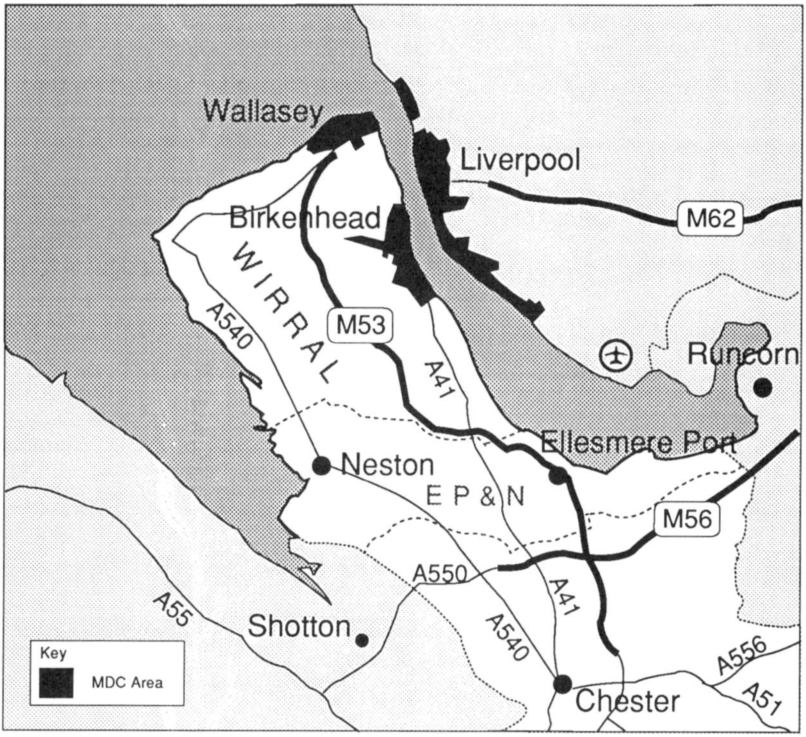

Fig 4.3 The Merseyside Development Corporation designated area

Case Study 18
Site assembly by Development Corporation

Development corporations are well placed to overcome ownership constraints to enable new development and urban regeneration to take place.

This 2.85 hectare site was assembled by the MDC in 1988. The bulk of it had previously been owned by the Mersey Docks & Harbour Company. This land was at one time in port related use but the nearby docks were no longer operational and the surrounding area had gone into decline. Both land and buildings had been let over the years to a variety of temporary uses and buildings had fallen into disrepair. In addition the existing infrastructure was outdated and access poor...

...Nevertheless it was unusual for the MDHC to release land other than for port related uses. In this particular case, however, no such further use was anticipated and the area required a massive injection of capital.

Since it was set up MDC had been hoping to acquire this site. After protracted negotiations involving exchange of another site, acquisition was completed. A larger area of adjacent land was intended to be the site of a prestigious residential and commercial development known as 'Twelve Quays', and it was anticipated that complementary development would take place on the smaller site discussed here.

4.3.3 Private development interests

Speculative developers were far less important in the Cheshire—Wirral corridor than either manufacturing companies or the public sector. This reflects the lower returns which industrial development, particularly in such regions as the north-west, are thought to provide in comparison with retail and office development in the south-east, for example. Some commentators (Morgan, 1990) have argued that extensive public sector provision of industrial accommodation depresses rental levels and deters private sector provision. There was little evidence, however, in the Cheshire—Wirral corridor to suggest that reduced public sector activity would lead to a rapid increase in speculative development by the private sector.

Most speculative industrial development in the Cheshire—Wirral corridor was undertaken by local or regionally based companies. There was only limited involvement by nationally known names. In contrast to their reluctance to undertake traditional industrial development, some major development companies showed particular interest in B1 development, notably at Crewe and Chester business parks. The Chester business park, initially promoted by McAlpine, has proven extremely popular with prestige companies seeking very high quality sites. Such development, however, is of very little interest to traditional users. Only locally and regionally based development companies appeared sufficiently aware of market opportunities in the Cheshire—Wirral corridor or prepared to take the risk of industrial development here (Case Studies 19 and 20).

Case Studies 19 and 20
Speculative initiatives by local developer

The market perspective of local, regional and national development companies will differ, as will the extent of their awareness of, and confidence in, local opportunities. Smaller, locally based companies play a key role in market niches which would be less attractive to other development interests.

These two schemes located within the older part of Crewe, and promoted by a local development company share many common features. In each case the site comprised an existing building, which was refurbished and subdivided, and an area of vacant land where new industrial units, typically ranging from 100 square metres to 300 square metres were later constructed. Development was phased over a number of years, in part reflecting the flexible working arrangement of the company. This company has both a property division, which acquires sites, and a building division which not only works on company sites but also on outside contracts.

Site 19 was acquired by the company in 1983. The existing building was subdivided into four units and a showroom. Subsequently, seven further units of approximately 300 square metres were constructed in two phases on the vacant land to the rear. The whole development was completed by 1987 at which time units were let at £26 per square metre. Flexible terms were offered to tenants, from twenty-one year leases, with three year rent reviews, to three year leases.

Site 20, which was acquired in 1986, involved the refurbishment and subdivision of an existing building providing 1400 square metres of industrial and office space, and the construction of eight new workshops, of 100 square metres each, on the remainder of the site. These two phases were completed in 1988 and 1989 respectively and the whole development was successfully let to small businesses, again on flexible leases.

In total 30 units were thus provided, half of which were new build. Local estate agents felt that these schemes helped fill a gap in the market by providing industrial space for small firms who could not afford rental levels on the new industrial estates in Crewe, at a time when there was a shortage of small starter or nursery units.

Successful speculative development requires not only detailed knowledge of local market conditions but also a careful assessment of likely future trends in rental levels. The future, however, is inevitably uncertain and development plans may be abandoned when anticipated growth in rental levels is insufficient to justify investment (Case Study 21). Furthermore, the strength of demand varied within the Cheshire—Wirral corridor, as evidenced by differentials in rental levels and land values, and the slower response to economic recovery in Wirral. Thus while local or regional development companies became more active in Crewe and Chester from 1987, they made only limited impact in Wirral until 1989. This, however, was the height of the boom. The marked increase in interest rates in 1990 and economic recession, inevitably undermined new confidence in the local market and again postponed the proposed speculative development of substantial sites (Case Study 22).

Case Study 21
Rental growth proves insufficient to justify development

As the economy came out of recession in the late 1980s, the refurbishment of obsolete industrial buildings was an important feature of development activity in traditional industrial areas. Project viability, however, depends not only on estimated costs, but also on the realization of anticipated rental growth. A re-appraisal of this scheme's viability 18 months after the site had been acquired resulted in the planned refurbishment of existing buildings having to be abandoned.

This site of 0.6 hectares is located in an area of Wirral, where the development market was characterized by small scale renovation and redevelopment of low-cost industrial property. A vacant warehouse of 4500 square metres covers almost the entire site. The building which dates from the early 1960s, is structurally sound, but has been vandalized. The previous owner vacated the building in 1985, moving its activities from this cramped site to a regional distribution depot elsewhere in Wirral. The property was subsequently marketed for 18 months at £400 000.

In July 1988, after several months of negotiations, the property was acquired by a private company for £175 000. This company, a family concern, had a steel stockholding division based in Liverpool and a property development division, specializing in the refurbishment of older industrial buildings. Planning permission was obtained in...

...January 1989 to convert the warehouse into 50 small workshops and offices. The company estimated that the refurbishment and conversion of the warehouse would cost between £200 000 and £300 000 and that a minimum rental level of £3.75 per square foot would be needed to make the scheme commercially viable.

When the building was acquired in 1988, new units in the locality were letting for up to £2.60 per square foot and rents were rising as confidence returned to the property market. Although rental levels continued to increase steadily over the next two years, a rapid rise to £3.75 per square foot, which the company had anticipated, never materialized. In 1990 therefore, the company abandoned the refurbishment and conversion plans and put the warehouse back on the market.

Case Study 22
The impact of the recession

The critical relationship between speculative development activity, the land market and wider economic events are evident in this case study.

This is a 10 hectare riverside site in south Wirral, last used as a power station in the late 1970s. The site was subsequently sold by the Central Electricity Generating Board to Merseyside County Council for reclamation purposes. After the abolition of the county council in 1986, the site passed from the Merseyside Residuary Body to Wirral Metropolitan Borough Council, which was responsible for completing the process of reclamation. A total of £2.1 million was paid in derelict land grant to reclaim the site. By the end of 1989, all buildings had been demolished and contamination eliminated. Drainage and access was still considered to be poor but it was anticipated that they would be resolved in the course of development.

Marketing of the site was undertaken in the autumn of 1989 and four bids were received from prospective developers. Three of these bids were in the region of £0.75 million for the freehold. The fourth bid which sought to acquire a 999 year ground lease, proposed a complicated financial arrangement, under which the local authority would receive a smaller initial capital sum but further receipts as and when phases of the proposed development were completed. The overall value of...

> ...this bid was estimated to be worth £1.8 million. Crucially, the developer owned an adjoining site, without which it may not be possible to achieve satisfactory access into this site.
>
> The local authority was pleased with both the financial arrangements and with the opportunity to retain the freehold, which the £1.8 million bid presented. The bid was accepted in April 1990 and contracts were ready to be exchanged by the middle of that year. Outline planning permission had already been granted for B1, B2 and B8 use and development was expected to commence by the end of 1990. A mixture of advance units and bespoke buildings was proposed. As interest rates rose sharply in late 1990 and demand fell away, the developer became increasingly reluctant to take on the commitment and contracts were temporarily put on ice until confidence improved. As a result of the recession, no further progress has been made. According to the local authority, there is little point in re-marketing the site until the economy improves.

4.4 SUMMARY

Between 1985 and 1987, the key characteristics of industrial land and property markets both nationally and in the north-west were an excess supply of developable land and a depressed market for existing floorspace. Regionally, London and the south-east were the most buoyant areas but even there it was not until 1987 that there began to be cautious optimism about the industrial market. In a buoyant economy in 1988, increased demand and a shortage of available sites in many areas led to rental growth and increases in land values. Location, however, was important and according to IRVO reports rental growth in the north-west was relatively low. There was still a plentiful supply of vacant floorspace in Liverpool, for example. The market did not improve significantly in the north-west until 1989. There was increased activity in the property market and strong demand for industrial land, particularly where there was motorway access.

Data specifically for the Cheshire—Wirral corridor broadly reflects this wider pattern, and in particular it highlights the relationship between industrial land and property markets. The impact of the economic recovery was reflected in the industrial property market as manufacturing companies sought to expand their activities. Shortages arose in particular submarkets in the Cheshire—Wirral corridor in 1988 and 1989, causing rents to rise towards the level necessary to justify new construction. In addition to a shortage of smaller starter and nursery units, INWARD, the agency responsible for attracting overseas investment to the north-west, also reported an unsatisfied demand for good quality premises over 1000 square metres.

As shortages in some sectors of the market began to be evident, demand by end users was transferred to the land market causing land values to rise, while higher rental levels for industrial units also made some new speculative development commercially attractive. Thus rental growth was slow between 1985 and 1987 but had increased significantly by 1989/90. At this point, land values also showed a sharp increase, although industrial land values in Chester had been increasing prior to this, in part as a result of supply constraints there. Increasing interest rates in 1990 and the subsequent downturn in economic activity resulted in rental levels and land values either levelling off or declining.

Manufacturing companies, rather than developers and investors were responsible for most of the demand for industrial land. Demand from speculative developers tended to be restricted to local or regionally based development companies. The evidence suggests that with the exception of interest in B1 space on prestige business parks, a period of sustained rental growth would be necessary to generate increased confidence in speculative development activity in the Cheshire—Wirral corridor. Alternatively, partnership between private developers and the local authorities, other public sector organizations or with landowners might well be necessary to make schemes viable in the future, and examples of such partnership will be given in later chapters. The Merseyside Development Corporation has been seen to play an increasing role in site assembly and land reclamation in Wirral, while in Ellesmere Port and Wirral, both English Estates and the local authorities already contributed significantly to the development of advance units to let.

Since demand for industrial land is clearly highly cyclical, responding to short term changes in economic activity, supply needs to be made available according to the timing set by demand. The next chapter therefore examines the aggregate supply of land and, in particular, the extent to which it meets the expansion requirements of individual companies. The following chapters then explore the development process in depth, investigating the extent to which supply constraints impede the market availability of land.

5

THE SUPPLY OF INDUSTRIAL LAND

The supply of land for industrial development has become a matter of increasing concern to many local authorities. This was intensified by the Town and Country Planning (Use Classes) Order 1987 which, for the first time, combined most office, light industrial and research and development activities into a single use class. Many authorities have also witnessed pressure to redevelop traditional industrial land for shopping or residential use. Industry itself is now seeking better quality locations, both in environmental and transportation terms, and inward investment is unlikely to be attracted by allocating poor quality sites for industrial development.

Local authorities have tried hard to respond to these pressures. Much of the experience gained in assessing housing land availability in the last decade is now being applied to industrial land. Through the process of structure plan revision, Cheshire County Council, for example, recently sought to ensure a minimum of five years' supply of industrial land with two years' supply immediately available. The relevant policy was deleted, however, by the Secretary of State in his modifications. A similar approach has been adopted by other authorities. In the West Midlands, for example, four market sectors have been identified, in each of which it is intended to ensure that sufficient land is always available.

To be readily available for industrial development, sites should be free, or easily freed, from supply constraints. The main supply constraints can be summarized as planning restrictions, physical or infrastructural difficulties, passive ownership and uncertainties about land values. Since supply constraints may respond to changes in demand, they are rarely absolute. A sharp increase in the demand for industrial land may not only encourage local authorities to remove planning restrictions but may also provide the necessary resources to overcome physical and infrastructural constraints, or tempt passive owners to sell at prices which they consider to be attractive. Relatively short term supply constraints should be distinguished from land shortages, which can be attributed to a longer term imbalance between supply and demand. Such an imbalance can usually be traced to market imperfections or even market failure. Poor exchange of market information between buyers and sellers would be a good example of this.

This chapter investigates the aggregate supply of industrial land in the Cheshire—Wirral corridor between 1985 and 1989 and also the extent to which it satisfied the particular needs of companies wishing to expand. Later chapters then examine the existence and severity of supply constraints to industrial development and explore the processes through which such constraints are most likely to be overcome by different development interests.

5.1 AGGREGATE SUPPLY

The research identified, for the four districts in the Cheshire—Wirral corridor, all sites of 0.2 hectares or above either developed for manufacturing industry between 1985 and 1989 or, if remaining undeveloped at the end of 1989, considered by the local planning authorities to have potential for industrial development. Of course, some of the latter land may eventually be developed for offices or research and development rather than traditional industrial uses but if the possibility of industrial development existed, these sites were included. Take up between 1985 and 1989 was recorded at 164 hectares. Potential supply at the end of 1989 was 708 hectares.

Take up and potential supply have been disaggregated according to the four distinct types of area defined in Chapter 2 (Table 5.1). The inner urban area accounted for only approximately 8% of take up between 1985 and 1989 and 14% of potential supply at the end of 1989. Throughout the corridor the rate of development tended to increase in the later years, as the economy improved. The potential supply of industrial land at the end of 1989 would last a further 22 years, at mean annual take up rates between 1986 and 1989. However, this aggregate figure is extremely misleading. It takes no account of market segmentation by size and location. Furthermore, not all sites included within potential supply were necessarily free from physical and infrastructural constraints, or in receipt of planning permission or available to purchase. In the Cheshire—Wirral corridor, the quality of potential supply was at least as important as the overall quantity and is therefore investigated in detail in the following chapters. In the meantime, however, attention is concentrated on the supply of land for expansionary development either through the intensification of use of existing sites or by new development on adjacent land.

Table 5.1 Overall take up of industrial land 1985—9 and potential supply at end of 1989

	Inner Urban		*Major Urban*		*Minor Urban*		*Rural*		*Total*	
	Sites	*Ha*	*Sites*	*Ha*	*Sites*	*Ha*	*Sites*	*Ha*	*Sites*	*Ha*
Take up										
Completed in 1985	1	0.2	-	-	-	-	-	-	1	0.2
Completed in 1986	3	0.7	10	5.1	2	0.6	2	0.3	17	6.7
Completed in 1987	1	0.4	10	6.4	2	1.1	3	2.2	16	10.1
Completed in 1988	1	0.2	8	9.8	1	0.4	3	6.6	13	17.0
Completed in 1989	8	3.7	16	13.4	1	0.1	3	1.8	28	19.0
Under construction 12/89	8	7.1	10	17.2	4	2.5	4	4.6	26	31.4
Phased estate development	-	-	6	76.3	2	3.0	-	-	8	79.3
Total take up	22	12.3	60	128.2	12	7.7	15	15.5	109	163.7
Potential supply	59	95.7	94	485.8	15	61.3	25	65.5	193	708.3
Total take up plus potential supply	81	108.0	154	614.0	27	69.0	40	81.0	302	872.0

5.2 THE SUPPLY OF LAND FOR EXPANSIONARY DEVELOPMENT

Particular concern has been expressed about the supply of land immediately adjacent to existing firms, or within their existing sites, on which expansionary development can take place. For instance, Fothergill and Gudgin (1982), have suggested a relationship between supply constraints upon land in urban areas and industrial decentralization to small towns and rural areas. Specifically, they contend that:

> *The available evidence suggests that this slower growth* (of employment within established urban areas) *is probably due to a higher proportion of city-based firms finding themselves in locations where they are unable to expand because their factories are hemmed-in by existing urban development. As machinery displaces*

> *more and more labour on the shop floor, employment in the city falls, while the lion's share of any growth in employment goes to smaller towns where there is more room for factory extensions and new premises* (p.112).

In aggregate, it will be true that there is a greater preponderance of greenfield sites, unfettered by existing urban structure, in smaller freestanding towns and rural areas than in traditional industrial locations in urban areas. The case study evidence discussed below suggests, however, that these spatial distinctions are an imprecise tool with which to describe the supply constraints faced by individual manufacturing interests seeking to expand.

There are clearly two distinct dimensions to the supply of land for expansion. One is the extent to which firms own sufficient land, often within the boundaries of their existing sites — that is **owner-specific** land. The second is the extent to which land is available adjacent to their existing sites which may be acquired to facilitate expansionary development. Each of these will be considered separately as they bring into play different market processes, strategies and constraints.

5.2.1 Owner-specific expansion land

The presumption often made in spatial explanations of changing patterns of employment (the 'urban—rural shift' debate) is that in many older urban areas existing firms are accommodated in cramped premises with little or no potential for further in situ development. There are, indeed, such sites to be found in the inner urban area of the Cheshire—Wirral corridor (Case Study 23) and these can be contrasted with those companies in rural areas with plentiful reserves of land for expansion (Case Study 24).

This picture, however, is an over simplification. It is evident, for example, that even within already cramped sites in urban areas, it may be possible to meet changing production requirements by recycling land (Case Studies 25 and 26). Indeed this strategy may be adopted even where companies own additional adjacent land. In Crewe, for example, the 24 hectare site owned by Rolls Royce Motors is now almost fully developed. Although the company also owns over 2 hectares nearby, it is anticipated that future development will take the form of the replacement or improvement of existing buildings. Thus, in 1989, when the construction of a new 6000 square metre paint shop was completed, the previous paint shop was gutted and used for car assembly (Case Study 43).

Case Study 23
Closure of Victorian premises

The experience described in this case study will be recognized by many as typical of long established companies located in the inner city.

In early 1990, a factory in Wirral, owned by a nationally known food manufacturing company, closed. Although the premises had been owned by the company since the 1930s, the buildings themselves were approximately one hundred years old. They were vermin infested and foul drainage had virtually reached capacity. Almost the whole of the site was built up, with no room left to expand and access was poor.

In 1989, the local authority had been prepared to make an adjacent site of 0.5 hectares available to the company to secure its long term future at this location. However, when the company merged with another operator, it decided to concentrate production at a more modern and spacious plant in Warrington. Around 150 jobs were thus lost.

The freehold of this site was marketed in April 1990 for £700 000. Unless the premises could be converted to smaller industrial units, it was highly unlikely that the site would ever again support the previous densities of employment.

Case Study 24
Ease of rural expansion

In contrast with Case Study 23, this is a classic example of the relative ease of expansion in a rural location. The 8 hectare site formed part of a former airfield in a rural area north-west of Nantwich. It had been owned freehold by a manufacturer of timber packaging since 1966, when the company was first formed, and had been developed progressively as required. In 1989, two developments were under construction, one an extension and the other a new building. Each provided around 1200 square metres of additional floorspace. Approximately one third of the site has further development potential. By the acquisition of this large rural site many years ago, this well established local firm was able to expand in situ and could accommodate its need for large areas of land for open storage.

Case Study 25
Expanding in a tight corner

This case study, together with the one below, shows how companies may meet their changing production requirements by redeveloping land within their existing sites.

This site of 0.27 hectares in the inner area of Wirral was acquired from the local authority by a small manufacturing company in 1966. In 1988, the company applied for planning permission to extend its existing warehouse and offices.

The layout had to take account of the tightness of the site. While a vacant yard could be used for the construction of the office extension, the warehouse extension had to be built on an existing car park. The local planning authority, concerned at the potential loss of car parking, required the layout to be altered to maximize car parking on the remainder of the site. The site was then fully developed and there was no adjacent land on which it might be possible to expand. The company, however, did not plan to expand further in the next five years. If necessary in the long term, the company would first seek to acquire land in the locality. Only if it were unable to do so, would it consider a move outside the borough.

Case Study 26
Demolition for expansion

The British subsidiary of a multinational oil company had occupied this 10 hectare site in Wirral docks on a long term ground lease since the 1890s. The site is now almost fully built up. In 1988 and 1989, the company redeveloped two parts of the site in order to provide a 2500 square metre warehouse for heavy storage and a further 2000 square metre warehouse for light packing. Existing buildings had to be cleared on both plots to enable these developments to proceed.

Conversely, turning attention away from the urban areas, it is possible to point to long established sites in small towns and villages which, like many sites in more urban locations, suffer the constraints imposed by restricted access and surrounding development, and which offer little space for in situ expansion, the sites now being almost fully developed. One such example is

Fig. 5.1 Narrow cobbled access to village site (Case Study 75)

a site located in a conservation area in a small village near Chester. It has been owned by the same company since the 1830s. Access from the junction of two village streets is via a narrow cobbled road which must be used by heavy lorries. The company wished to expand, diversify and retain its existing workforce, but local opportunities to relocate are constrained by the green belt (Case Study 75).

Many of the relevant case studies show that the land strategies and historic patterns of growth of individual companies are of greater importance than spatial location, in understanding the supply of land for expansion within existing sites. The ability to expand at any moment is clearly influenced by the extent to which the company, in its initial location decision, acquired sufficient land to enable future growth or only to meet short term production requirements (Case Study 10), and by the extent to which the site has been developed over the years. Even within inner urban and major urban areas, there are examples of large historic landholdings by major manufacturing companies which have enabled them to undertake continuing and recent development with least disruption to the manufacturing process (Case Studies 27 and 28). Although it is unusual for minor manufacturing companies to retain sufficient land to meet long term production objectives, some examples of this can also be found (Case Study 29).

Case Study 27
Future expansion secured by landholdings

This site of 10.6 hectares had been occupied by a manufacturer of food additives since 1982. It was held on a 125 year ground lease from the parent company and was part of the latter's extensive landholdings in Wirral.

The site had originally been developed in the 1950s by another subsidiary of the parent company. This venture was not a commercial success and the existing buildings had become redundant by 1980. In 1982 the site was cleared for redevelopment. The redevelopment costs were in the region of £10 million and the availability of assistance from the Department of Trade and Industry was critical to the decision to invest here. An alternative greenfield site in the south-east had been under consideration.

There have been three expansionary developments on the site since 1985, the last of which was the replacement of an obsolete laboratory and sales building by a new development of 1520 square metres. The availability of land ensured that the obsolete building could remain operational while the larger new building was constructed adjacent to it.

Only 30% of the site had been developed by 1990. Although it was evident from the company's expansion plans that the whole site would not be fully utilized for many years, surplus land was to be retained to meet future needs. The absence of any pressure to release this land to other potential users reflected the land management strategy of the parent company and the leaseholder's own financial and production plans.

Case Study 28
Continuing growth on long established site

In 1971 a nationally known manufacturer of gas products relocated to Crewe from Stoke-on-Trent, having to find a larger site to enable it to expand. It acquired the current operational site of 8.9 hectares, together with adjacent land. A number of expansionary developments have taken place since 1985 within the boundary of the operational site, the largest of which was a cylinder receiving building of almost...

...1500 square metres, completed in 1990. Although the company's land requirements are particularly affected by the nature of its activity and the need for a safety zone at the site boundary, its landholdings enabled unimpeded expansion to take place. It is interesting to note that the company no longer owns adjacent land, having been encouraged to dispose of its surplus by the local authority.

Case Study 29
Long term strategy of minor manufacturing company

This is an unusual example of a minor manufacturing company acquiring sufficient land to allow future expansion.

In 1981 this electronics company was compelled to move to Crewe from Nantwich as a result of the shortage of available sites to enable it to expand there. It was anticipated in 1981, that the new site would meet the company's requirements for 20 years, during which time the site would become fully developed. Indeed, in 1987 a warehouse extension of just under 1000 square metres was completed and there was potential for the development of a further 2000 square metres. The acquisition of a site sufficient to meet its long term requirements thus enabled the in situ expansion of this company.

Finally, it is important to note that despite the traditional supply constraints in inner urban areas which have given rise to spatial explanations of the urban—rural shift, there are recent examples even in such locations as Wirral docklands of the acquisition of sites for new development with sufficient land to allow for projected future expansion (Case Study 16). The public sector (whether the local authority, English Estates or development corporation) plays an important role through land assembly and reclamation in changing the pattern of land availability in traditional industrial areas, bringing forward sites in a range of sizes to meet both the immediate and longer term needs of manufacturing interests. The role of the public sector and its influence on land supply in inner urban areas may in future challenge spatial explanations of a continuing urban—rural shift.

In summary, in looking at the potential for expansion on owner-specific land in different types of area, a more diverse pattern emerges than that suggested by earlier research. The critical determinants are the past strategies and behaviour of individual companies. It might be argued that past patterns of development which located industry in what is now the heart of the urban area will suggest that a higher proportion of companies there will have very little land available within their sites. The problem, however, is clearly not exclusive to inner urban areas and indeed, even here, some major manufacturing companies are still able to expand, either by virtue of the scale of their landholdings or through redevelopment. Furthermore, as the public sector continues to address the problems of dereliction and land vacancy, it may increasingly be possible for companies to locate here and grow and for existing companies to expand.

5.2.2 Expansion onto adjacent land

In order to expand onto adjacent land, which is not already in their ownership, it will be necessary for manufacturing companies to enter the market, and supply constraints will become important. As above, it is not possible to make any categoric statements on the relevance of the spatial distribution of supply constraints on the market availability of expansion land. Those who would explain the urban—rural shift in this way would perhaps single out the example of a site in Ellesmere Port (Case Study 30). This site and the surrounding area are fully developed and the company was therefore beginning to consider the advantages or otherwise of relocating out of the district. Examples can also be given of the problems in urban areas of assembling adjacent land in fragmented ownership, of overcoming poor ground conditions and the cost of refurbishing existing obsolete buildings thus acquired (Case Study 31).

Case Study 30
Future dilemma for company on constrained urban site

Over half of the 7 hectare site in Ellesmere Port owned by a manufacturer of specialist sheet steel was covered by buildings. The remainder was necessary for open storage and no further development was therefore possible without the acquisition of adjacent land, which was also developed. Furthermore, there were no suitable sites of the right size in the area to which the company might relocate. In 1990 the company was beginning to explore the possible advantages of...

...relocating to Deeside, or even to Humberside which would afford access to European markets.

Case Study 31
Site assembly to facilitate expansion

This case study concerns the expansion onto adjacent land of an international chemical company, established in Wirral since 1964. The case study demonstrates the importance of available expansion land to an established company which had considerable investment in its existing plant and was unable to operate efficiently on a split site basis. The company was therefore willing to overcome the physical constraints on an adjacent site which might have deterred new development by any other party.

The original site of approximately 24 hectares, had become extremely congested. The need was identified for a new tanker marshalling yard and for the relocation of a car park, which together would release land within the main site for future operational use.

The relocation of an adjacent engineering company to Chester provided the opportunity to acquire expansion land freehold. This site was not in itself large enough to accommodate the marshalling yard. An equivalent area was therefore acquired on a 125 year lease from a major manufacturing company with large landholdings in the locality. It took 15 months from the first expression of interest in the adjacent engineering company's site to conclude the assembly of the expansionary site.

Although poor ground conditions constrained site layout and added to the overall costs, they did not inhibit development. One third of the site had previously been tipped and could therefore not take a heavy load. The installation of surface drainage at a cost of £200 000, and the renovation of an existing building at a cost of £100 000, also contributed to the overall development costs of £1 million.

It is equally possible, however, to illustrate the impact of supply constraints in rural areas. Planning constraints, for example, prevented the expansion of a long established local company in one Cheshire village (Case Study 79), while valuation disputes prevented the expansion of another company onto land adjacent to its site in the green belt. This opportunity had arisen because the site had been designated an inset in the green belt in the Chester Rural Areas Local Plan (1985), specifically to enable the expansion of this company. Negotiations subsequently broke down (Case Study 96). Problems of land availability adjacent to companies' existing premises can be found even on the urban fringe (Case Studies 32 and 33) and there will be many cases throughout this text which illustrate the existence of physical and infrastructure constraints in rural areas, which are no less problematic than in urban locations. Indeed, one such example concerned not expansion onto adjacent land, but intensification of site use by a company manufacturing dairy products. It became necessary to spend £250 000 on an effluent treatment plant (Case Study 34).

Case Studies 32 and 33
Two neighbouring companies in land trade

Browns Ltd, a major manufacturing company with interests in Europe and the Far East, built a factory in Wirral in 1970. Operations expanded successfully and by the mid 1980s, the site had reached maximum capacity.

The neighbouring company, Smiths Ltd, had acquired a large greenfield site in 1966, sufficient to satisfy their expansion requirements for many years to come (Site 33). In 1986, Smiths were approached by Browns, who wished to acquire under utilized land to the rear of Brown's factory (Site 32). Although the land had outline approval for industrial use, part of it was used as a car park and the rest for tipping industrial waste.

Smiths were willing to negotiate with their neighbour but the deal took 12—18 months to conclude. Smiths were only prepared to release the expansion land to Browns if it could be replaced by a site of comparable size, adjoining their own boundary. The greenfield site identified was in local authority ownership. Furthermore, Smiths would not acquire the replacement site unless it had planning approval for industrial use. This was refused, as a result of public pressure. Permission was given, however, at the second attempt primarily because of the implications for the local workforce if Browns were forced to relocate. Not only would 60 existing jobs disappear but a further 60 potential jobs would be lost.

Once Browns had acquired Site 32 from Smiths they had to dig out...

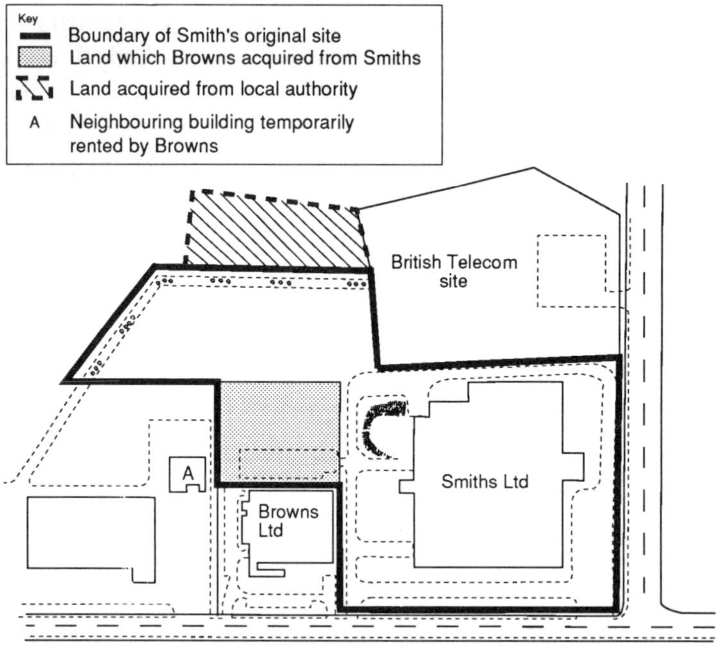

Fig. 5.2 Site plan for case studies 32 and 33

...existing fill, remove it off site and replace it with properly compacted material, prior to development. In the meantime, Browns had to rent a nearby building for 18 months, as an overflow.

The financial compensation which Browns paid to Smiths for the transfer of the poor quality backland site further illustrates the great pressure upon Browns to secure the land for their expansion. Rather than simply pay the value of the land transferred, Browns agreed to compensate Smiths for their full acquisition costs of the replacement greenfield site (acquired at higher value from the local authority) and provide an additional sum as compensation to Smiths for the loss of their car parking. Browns did, however, receive derelict land grant and regional development grant from central government which proved critical in persuading them to tackle the physical and infrastructural constraints evident at Site 32, rather than to relocate elsewhere.

Case Study 34
Constraints on expansionary rural development

Food processing is an important manufacturing activity in Cheshire, and this case study demonstrates very effectively some of the physical constraints on industrial development which may arise in rural areas. In 1983 a new company was created by local farmers and businessmen to manufacture dairy products on this 2 hectare site in a rural part of Chester district. Initial development of the site presented few problems. There were a number of constraints, however, on the intensification of use of the site particularly related to access and infrastructure.

Later phases of development necessitated the construction of an effluent treatment plant on site at a cost of £0.25 million. Manufacturing processes produced a considerable amount of waste which had previously been discharged into a stream. The plant was constructed in 1988/89 in response to the possibility of prosecution. Secondly, lorry access was poor since the site was approximately one and a half miles away from the main highway, on a minor road. This had implications both for the company and for the local community. The problems were alleviated, both by the landowner providing land to the local authority for the construction of passing places and by prohibiting lorries from turning left out of the site in order to keep them out of the village. Finally there were objections from neighbouring landowners, which delayed the most recent development by a year.

Despite these various problems, however, the site is now fully developed and, by producing a reputedly good product, the company's market extends to Grimsby, Carlisle, Birmingham and Holyhead. The company now has a second site south of Chester, an annual turnover of £30 million and employs around 120 people in total.

5.2.3 The take up of available expansion land

Even where expansion land is available to meet a company's immediate production requirements, its development will depend on wider and longer term considerations. Where sites are beginning to be congested and existing buildings are outmoded, for example, whether in urban or rural settings, companies will weigh carefully the potential long term advantages of relocating altogether (Case Study 35). Furthermore, financial constraints may inhibit development and lead companies to resolve their needs more cheaply in the short term by renting property (Case Study 36).

Case Study 35
Expansion plans now under review

This site had been in industrial use for 80 years and had produced textiles for 40 years. Planning permission was given in 1988 for an extension of over 1000 square metres. Although in 1990, the additional space was still required, the proposal had not been implemented and there was no likelihood that it would be undertaken in the foreseeable future. This local company is a subsidiary of another company, following a take over in 1973. Although the subsidiary company was profitable in 1990, the parent company was no longer buoyant. Indeed, the recession had forced a number of its competitors out of business. The proposed expansion was therefore prevented in the short term by the availability of development finance at a corporate level.

However, when this long established industrial site near Nantwich town centre was first developed, it was surrounded by open fields. Now it is largely enclosed by residential development, built from Victorian times onwards. As a result, the area of land which remained undeveloped within the 0.7 hectare site had poor access to the public highway. Access to the main road was via a narrow residential street used by service vehicles and for car parking. Furthermore, visibility at the junction with the main road was poor. The managing director now felt that whether the expansion in fact took place in the future would depend on a re-evaluation of the advantages of moving to a new and larger site.

Case Study 36
Expansion of a branch plant restricted by internal controls

This site of 14.8 hectares, north-east of Neston town centre, had been leased by Cheshire County Council to a manufacturer of armaments since 1980, having previously been let to the parent...

...company for electrical engineering. In recent years there has been further development of the site. In particular an extension to an existing storage and warehouse facility was completed in 1987 and provided over 1100 square metres of new floorspace. The availability of land on this large site had therefore afforded the possibility of expansion. Indeed half of the total site area remained undeveloped.

It is interesting to note that factors other than physical constraints on land availability have inhibited further development. The main constraints were internal financial controls and the decisions taken by the head office of this multiplant company, which also has bases in Scotland and on the south coast. A new office development of 1200 square metres to replace cramped temporary accommodation on the site was given planning permission in 1988 but was subsequently blocked internally because of the estimated £1 million cost of construction. Similarly, despite the availability of expansion land within the site, because of short term considerations of cost and timescale, the company rented premises on a five year lease on the adjacent industrial estate. Companies may thus satisfy their space requirements without undertaking new development on available land.

It will be evident in both the above cases that development by local branches was constrained by the wider perspective of the parent company or head office. Local land management and development strategies are determined in the case of multinational and multisite companies by wider corporate considerations. While one solution to changing production requirements may be to rent property in the short term, at the other end of the spectrum, such companies may take over smaller firms with compatible production space (Case Study 37).

As a result of these several strategies, and the long term retention of land for future expansion discussed earlier, large areas of potential employment land can remain undeveloped for many years. Although an opportunity cost may attach to holding reserves of vacant land, the direct costs are little and it may be that expansion land will be held for many years before there is any re-appraisal of the company's long term needs. Indeed it may not be released in some cases without either corporate restructuring or public sector intervention. These issues will be returned to in Chapters 6 and 9.

This section will conclude, however, by looking briefly, not at corporate inertia, but at the diverse interests which manufacturing companies may have and how even potential expansion land may be viewed and managed as a financial asset (Case Studies 38—40). Manufacturing companies may acquire land as a speculative investment. Equally they will measure the immediate monetary return from disposing of potential expansion land against the long term production flexibility which such landholding offers.

Case Study 37
The impact of business strategy on land requirements

This is an example of a site in a major urban area where there was sufficient room for expansion. It demonstrates how in the case of multiplant and multinational companies, decisions to develop land at a local level are influenced by the company's broader business strategy.

The owner of this 0.6 hectare site in Crewe was a manufacturer of business stationary who relocated there in 1979 from Newcastle under Lyme, in order to expand. In 1986, planning permission was given for further development comprising a 1000 square metre office and factory extension on a residual area of 0.25 hectares. This, however, was not implemented. When permission was sought, the proposed development had been only one option open to the company, which was a subsidiary of an American owned group. The subsequent acquisition of two other companies by the group increased production capacity and removed the need to undertake expansionary development here.

Case Study 38
Adjacent expansion land marketed

In the late 1960s, a nationally known food manufacturer acquired this 3.5 hectare site in Wirral in order to rationalize activities from two smaller operations, one in Birkenhead and one in Bristol, into a single, purpose built factory. The site proved sufficiently large to allow gradual expansion to take place. The most recent developments comprised a high bay warehouse of around 1500 square metres and a loading bay and canopy of just under 1000 square metres. Apart from the need to pile foundations for the warehouse to a depth of 9 metres, both developments proved straightforward and were undertaken within a...

...year of planning approval.

After the completion of the warehouse and loading bay, only 0.2 hectares with development potential remained within the site curtilage. However, the company owned an additional 1.5 hectares adjacent to the site. This land had previously been tipped, making ground conditions problematic.

In 1986, the company attempted to market this site, but interest was expressed only by a scrap metal dealer, who the company considered would be an incompatible neighbour. The company then decided that the value of the 1.5 hectares, if sold, (about £50 000—60 000 at the time) was insufficient to justify jeopardizing the potential for future expansion of a plant in which several million pounds were invested.

Case Studies 39 and 40
Speculative acquisition of adjacent land

The behaviour of this company demonstrates a duality of interest. On the one hand it has developed its operational site to capacity to meet its requirements. It has also acquired adjacent land, but as an investment, with a view to selling it for a higher value. Case Study 39 was a 1 hectare site in Chester which had been occupied for over five years by an international company. The company first opened a branch in Chester in 1974, and relocated to premises on the present site in order to expand. Since 1985 there were two further developments on site. In 1986 garage space on the ground floor of a two storey building was converted to provide additional office and laboratory space, and in 1988 a new building providing 1000 square metres was completed. The site was then fully developed. Although the company owned the adjacent site (Case Study 40), further expansion here was unlikely. Future development by the company was likely to be abroad rather than in the Chester area.

The adjacent 1.4 hectares (Case Study 40) was acquired freehold from British Rail in 1986 for £35 000 for a long term investment. Although the site was allocated for industry and warehousing use and was adjacent to a railway line in a predominantly industrial area, there were houses nearby. The company hoped to realize a gain in value by marketing it for residential development, for which planning permission was sought.

5.2.4 The aggregate supply of expansion land

The many case studies in this chapter have shown how the ease or difficulty of expansion varied sharply from site to site, and the same constraints and strategies can be identified across the spectrum from rural to inner urban areas. A more refined analysis would be necessary, however, to conclude the discussion as to whether any clear difference existed from one type of area to another.

To examine this, information was collected on the type of development proposed in each type of area (Table 5.2). This analysis covered not only the 164 hectares either developed between 1985 and 1989 or under construction at the end of 1989, but also a further 174 hectares, for which development proposals had been made through planning applications or by other means. Throughout the Cheshire—Wirral corridor, new freestanding developments accounted for 75% of these 338 hectares of land proposed for industrial development. However, the share of expansionary development clearly rose as one moved outwards from inner urban and major urban areas to minor urban and rural areas. At an aggregate level, the relative availability of land for expansion between different areas may be considered to explain these variations.

To test this further, it was necessary to examine the expansion plans of individual companies. Of the 60 sample sites which had been developed between 1985 and 1989, 40 were owned and occupied by manufacturing companies. Each of these companies was asked whether it had any further development plans and whether it had identified suitable land on which to build. All 14 companies which planned further development already owned suitable land either within their existing sites or adjacent to them. Seventeen of the remaining 26 companies who had no further development plans also owned suitable building land either within their existing sites or adjacent to them. Of the remaining nine companies not planning further development, six could identify suitable land elsewhere in the borough. In two of these cases the land was already owned by the company. This left only three companies (none of whom planned to expand) who were not able to identify any suitable land for expansion either within their existing site, or adjacent to it, or elsewhere in the borough. In all these findings, there were no significant differences between inner urban, major urban, minor urban and rural areas.

Table 5.2 Type of development completed, under construction or proposed as at 31 December 1989

	Inner Urban		*Major Urban*		*Minor Urban*		*Rural*		*Total*	
	Ha	*%*	*Ha*	*%*	*Ha*	*%*	*Ha*	*%*	*Ha*	*%*
New freestanding development	36.0	78	188.0	81	11.7	49	17.7	50	253.4	75
Expansionary development:										
— within existing site	2.8	6	11.7	5	7.5	31	12.8	37	34.8	10
— on the adjacent site	2.6	6	24.4	10	2.9	12	1.3	4	31.2	9
Change of use/ subdivision and mixed developments	4.8	10	8.6	4	2.0	8	3.2	9	18.6	6
TOTALS	46.2	100	232.7	100	24.1	100	35.0	100	338.0	100

5.3 SUMMARY

Site by site analysis of the supply of land for expansionary development does not confirm the conclusions of earlier work based on aggregate supply analysis. It must be stressed, however, that the database excluded companies which had never made a planning application and excluded available expansion land as yet unknown to the local planning authorities. The specific characteristics of the Cheshire—Wirral corridor and of the study period, as well as the relatively small number of companies interviewed, also need to be borne in mind. It may be the case, for example, that the recession of the early 1980s had created sufficient spare capacity within existing factories to allow subsequent increases in production to take place without any need for additional floorspace. If so, the potential for on site expansion may not have been an important issue to many of the companies interviewed.

The research therefore cannot confirm, or refute, the extent to which site congestion and supply constraints explain the urban—rural shift of employment. While the hypothesis may seem a reasonable one, there are clearly other important contributory factors, such as the impact of economic restructuring, corporate restructuring and changing criteria in the location decision making process. There is evidence, for example, of spatial imbalances nationally in the growth of employment in high technology industry. Between 1981 and 1989, such employment declined in conurbations and more urbanized counties but increased in less urbanized counties by 7% and in rural counties by 12% (Keeble, 1992). This pattern will have little to do with physical pressures on existing sites, but will reflect instead the access and skills requirements of high technology industry, the residential location preferences of the workforce and the desire for good quality sites. The continuing impact of these wider economic phenomena, call for more detailed investigation.

Despite the necessary limitations of this study, what is very clear is that the ability of individual companies to expand depends on a combination of factors:

- Supply rigidities. There is evidence of the impact of planning constraints, physical constraints, valuation and ownership constraints throughout the local land market.
- Historic land management and development strategies. The extent to which past landholdings have anticipated future expansion needs, the level of existing development within sites and the ability of companies to redevelop parts of their existing site have been shown to be important determinants of the ease with which expansion takes place.

In particular, these findings suggest that the land management and development strategies of manufacturing companies, as well as those of other interests, need to be explored in much more detail, before supply issues can be resolved. An interest based approach to analysis may in fact be more helpful in examining land supply constraints on company expansion than a purely physically based approach. The next chapter therefore investigates the land management and development strategies pursued by various interests in order that the subsequent chapters can examine whether some interests are more successful than others in overcoming supply constraints and bringing land forward for development.

PART THREE

AGENCY RESPONSE TO SUPPLY CONSTRAINTS

6

LAND MANAGEMENT AND DEVELOPMENT STRATEGIES

Although the planning process may define the potential aggregate supply of land for industrial development, in practice the land which is brought forward will be the outcome of the several strategies of land ownership and development interests. This chapter seeks to classify these interests in order to provide a basis for exploring how different interests respond to the various supply constraints discussed in later chapters.

This part of the analysis is distinguished from earlier chapters in Part Two by the fact that the discussion will no longer focus on aggregate supply and demand data but will concentrate on the detailed evidence provided by the sample of 120 case studies, which have only been used thus far for illustrative purposes. These case studies enable a more refined classification of interests than that which was given in Chapter 4 on the basis of aggregate data. Consumers of land were identified there as being either production interests (whether major or minor companies), private developers or statutory agencies. In this chapter, not only will the role of historic 'landed property' interests and financial institutions be discussed briefly, but it will again become very evident that each category of interest may pursue more than one objective. Manufacturing companies, for example, may behave in the land market either as production interests or, where land is acquired or held primarily as an investment, as financial interests. Furthermore, as development strategies are implemented, or evolve in response to structural change, the behaviour of any agent may be transformed from that of a consumer to a supplier of land.

6.1 A CLASSIFICATION OF OWNERSHIP AND DEVELOPMENT INTERESTS

Land development has been defined as *'a purposive and premeditated material change in, or intensification of, the use of a parcel of land, brought about by an investment of labour and capital in land*' (McNamara, 1983, p. 88). Such development only takes place when a particular individual or organization is prepared to devote time, energy and resources to meeting a perceived need or demand for a development product. Models of the development process reveal its inherent complexity (Gore and Nicholson, 1991; Healey, 1991a). Demand may make development viable or need render it essential, but neither are sufficient to guarantee its delivery. Development feasibility, in site terms, must also be established. Development only becomes feasible when planning permission is obtained, physical and infrastructural constraints are overcome, and land passes to (or is already in) active ownership . When development is inhibited by planning restrictions, physical or infrastructural difficulties, passive ownership or uncertainties about land values, supply constraints can be said to exist.

During the 1980s, emphasis was placed in British land policy upon reducing supply constraints on land development. Measures were introduced to make the town and country planning system more responsive to the priorities of the development industry, as already discussed in Chapter 3. Much greater emphasis was given in derelict land reclamation to preparing sites for hard end uses, through tackling physical and infrastructural constraints. Passive land ownership in the public sector was identified through the introduction of public land registers. Local authorities were advised, when disposing of surplus sites, to avoid inappropriate methods of valuation which produced unrealistic asking prices (Valuation Liaison Group, 1985). Such initiatives, however, have never attained the status of a comprehensive or well integrated policy response to supply constraints on land development. Furthermore, a limited understanding of the motives and behaviour of the various actors in the development process has allowed these initiatives to be formulated and implemented, on the basis of some very questionable assumptions indeed. This is nowhere better illustrated than in the case of the policy response to passive land ownership.

Land ownership is increasingly recognized as a critical factor in the development process (Goodchild and Munton, 1985), but as Cameron, Monk and Pearce (1988) contend, such ownership influences are highly complex and extremely difficult to unravel. They suggest that:

> *Major owners of urban land appear to utilize their land holdings in a variety of ways. In both the public and private sectors, some seek to maximize revenue by active participation in the land market releasing and buying surplus land as appropriate conditions arise. Others hold excess land as a measure of coping with future*

growth without the accompanying need for future land acquisition or relocation of production elsewhere. And others appear to have no explicit policy for their surplus land. There is a particular need to understand how private and public bodies invest in, maintain and exploit their land holdings (pp. 124—5).

Research has also drawn a clear distinction between active and passive behaviour in land ownership. Landowners may be defined as active in the development process if they develop their own land, enter into joint venture development or market their land for others to develop. In contrast, passive landowners may respond to offers from potential developers (or fail to respond to such offers) but would otherwise hold on to land without development (Adams and May, 1991). However, despite extensive empirical evidence of widespread passive ownership in the private sector, policy has concentrated almost exclusively on passive ownership in the public sector.

Thus, for example, since 1980 central government has had the power, under the Local Government, Planning and Land Act, to release land in public ownership by issuing compulsory disposal notices. This power was exercised on 82 occasions between 1984 and 1990. In reality, and as shown by many of the case studies described here, local authorities are often very willing to sell land for employment development. Furthermore, public corporations such as British Rail have been increasingly active in reappraising their land holdings. Less attention, however, has been focussed by central government on the extent of passive land ownership in the private sector and, as will be discussed in Chapter 9, the powers which local authorities have to acquire land by compulsory purchase are both cumbersome and costly. It will clearly be important, in order to formulate appropriate policy responses to overcome supply constraints, that the actions, interests and strategies of the different interests which own land with potential for industrial development are better understood.

As a basis for empirical observation, ownership interests have been classified in relation to their function in the process of production. Massey and Catalano's (1978) threefold typology of former landed property, industrial land ownership and financial land ownership has therefore been adopted and supplemented by the inclusion of statutory land ownership. Of course, landowner behaviour may well vary within any one of these four categories (Adams and May, 1991). Different functions in the process of production may be associated with different behavioural characteristics, which result in different outcomes. The classification of ownership interests here, however, is intended to facilitate the analysis of behaviour and outcomes rather than to substantiate a particular theoretical approach. As this and subsequent chapters demonstrate, patterns can be discerned in the way in which land is brought forward for development, which relate significantly to the interests of those who own land.

As Chapter 2 explained, the research in the Cheshire—Wirral corridor investigated a sample of 120 sites in detail. Despite the much greater proportion of undeveloped sites in the population, it was decided, for comparative purposes, to select 60 sites which had been developed or were under construction and 60 sites which remained undeveloped. For the 60 sample sites developed or under construction, the research identified the relevant landowner both at the commencement of development and four years previously. Only 28 of these sites had been owned by the interest responsible for development four years before development commenced. For the 60 sample sites which remained undeveloped at the end of 1989, the research identified the relevant landowner both for that date and four years previously, at the end of 1985. Of these sites, 42 remained in the same ownership for the whole of the period 1985—9 (Table 6.1). The analysis therefore indicates that 18 of the 60 undeveloped sites had changed hands between the end of 1985 and the end of 1989 and that, of the 60 sites developed or under construction, 32 had changed hands in the four years prior to development. A total of 170 past and present ownership interests were thus identified, including the 50 previous site owners.

The data indicates that former landed property appeared to be of limited importance to the process of industrial development in the Cheshire—Wirral corridor. The Duchy of Lancaster, the Church Commissioners and a number of smaller landed trusts were included within this category. As might be expected, industrial land ownership was far more significant. Production interests have been subdivided into three categories. 'Major manufacturing' interests consisted of national and multinational companies and included such household names as Mobil, Rolls Royce and Vauxhall. 'Minor manufacturing' interests comprised the remaining regional and locally based companies. 'Other production' interests were primarily involved in transport, distribution and utilities. They also included two retailing companies and a small number of farmers.

Financial land ownership was subdivided according to whether the particular landowner was principally concerned with either investment or development. Three types of statutory land ownership were identified. Sites owned by the first type of statutory landowner, local authorities, were by far the most numerous. The second type, development agencies, consisted of English Estates and the Merseyside Development Corporation. The third type, other

Table 6.1 Ownership interests in sample sites

Ownership interest	*Undeveloped sites At end of 1989**		*Developed sites When development began**	
1. Former landed property	7	(7)	-	-
2. Production, major manufacturing	11	(9)	23	(17)
3. Production, minor manufacturing	2	(2)	19	(5)
4. Production, other**	9	(4)	2	(1)
5. Financial, investment	2	(2)	-	-
6. Financial, development	5	(1)	7	(1)
7. Statutory, local authority	16	(12)	4	(4)
8. Statutory, development agency	7	(4)	5	(-)
9. Statutory, other	1	(1)	-	-
TOTALS	60	(42)	60	(28)

* Figures in brackets indicate the number of sites which had been in the same ownership interest for at least the previous four years.

** Other production interests include agricultural landowners and private water transport companies, amongst others.

statutory owners contained only two organizations — the British Rail Property Board, referred to in Table 6.1, and an area health authority which released a site to a development interest.

The above classification will provide the context for later discussion of the extent to which each interest sought to overcome supply constraints to industrial land development. The rest of this chapter, however, explores the land management and development strategies of the various interests. It will illustrate the relationship between these strategies and the supply of land, as a basis for the more systematic exploration of ownership constraints in Chapter 9.

6.2 FORMER LANDED PROPERTY

'Former landed' interests were involved in the sample almost exclusively as the owners of undeveloped sites. Most of these sites were gradually being moved forward towards development. Time horizons were lengthy, however and progress often slow. Ownership behaviour tended to be passive. Most sites were released and developed only in response to approaches from developers and were not otherwise actively marketed (Case Study 41). Such land therefore tended to become available for development only at the height of the economic cycle. Otherwise, its availability was constrained or influenced by traditional land tenure, by perceptions of historic use, or by restrictive covenants. Thus one large and now important industrial site at Nantwich had been held in trust for the benefit of the people of the town since the 18th century and could only be released if exchanged for a site of comparable value (Case Study 113). Where former landed interests have been active in releasing land, the limited evidence suggests a very cautious approach to participation in the market (Case Study 42).

Case Study 41
Former landed property under no pressure to develop

This major site of 12.5 hectares south of Ellesmere Port was owned by the Church Commissioners. It had been in ecclesiastical ownership for 700 years and was part of much more extensive landholdings. It was allocated for housing development in 1961 when it was intended to be the location of an overspill estate for Liverpool. This never materialized and in the 1970s the site was re-allocated for employment use. The need for substantial investment in access and services, however, was well known. No particular action was taken by the Church Commissioners and the site remained in agricultural use until 1989. At that time, at the height of the property boom, the Commissioners were approached by a speculative developer, with proposals for a regional shopping centre of 50 000 square metres on neighbouring land in the Commissioners' ownership, which would help fund the infrastructure for a business park, again of 50 000 square metres, on this site. The location of the site, at Junction 10 of the M53, and within easy reach of Chester where land availability was constrained, made the proposed development highly attractive at the time.

Outline planning permission was granted in 1989, subject to a Section 52 Agreement to ensure the comprehensive development of both the shopping centre and the business park. The Commissioners then disposed of both sites to the developer in 1990, on a 150 year ground lease. The Commissioners' decision to sell was thus a response to broader economic processes, mediated through a development interest.

Case Study 42
Cautious marketing strategy by landed property interest

This site comprised a vacant farmhouse and outbuildings on approximately 0.4 hectares of land, in a rural area to the south-east of Crewe. It was part of a working farm until 1988, when the tenant farmer retired, at which time the landowner recognized the potential for commercial use, arising from the site's proximity to a new bypass. Planning permission was obtained for the conversion of the farmhouse to B1 use in 1988.

Feasibility studies indicated that repairs alone would cost in the region of £36 000, while comprehensive refurbishment to commercial standard would cost over £77 000 and would take a year to complete. The owner was reluctant to undertake this work without having first secured a prelet. The site was therefore marketed for 12 months until July 1990, when it was withdrawn. The failure to let the site can be attributed to the general downturn in the market, the lengthy lead in time for refurbishment and the desire of the freeholder to secure a single user for the entire 1000 square metres of floorspace.

As a result of the cautious marketing strategy, the proposed development which had the support of the Rural Enterprise Commission, was not implemented.

6.3 PRODUCTION INTERESTS

6.3.1 Major and minor manufacturing interests

Manufacturing companies accounted for 70% of all sample sites developed. As a result of the sample structure, this proportion is higher than in the population as a whole. A separate analysis of all industrial planning applications submitted in the Cheshire—Wirral corridor between 1985 and 1988 (apart from minor extensions under 1000 square metres) revealed that major and minor manufacturing companies nevertheless accounted for approximately 57% of all the applications submitted (Table 4.3).

One of the essential differences between major and minor manufacturing interests in the Cheshire—Wirral corridor appeared to be their approach to

Case Study 43
Land management strategy of a major manufacturer

The strategies of major manufacturing companies reflect a long term view of future land requirements.

Rolls Royce, the motor manufacturer, has occupied a 24 hectare site in Crewe since 1939. It has therefore been able to expand in situ over the years, to the point where this very large site, owned freehold by the company, is almost fully developed. The most substantial development in recent years was the construction of a new 6600 square metres paint shop, completed in 1989. The land within the curtilage of the operational area was then almost fully developed, but the company also owned approximately 2.5 hectares to the west, separated from the main site by a highway. This site could meet the company's future needs, but it was anticipated that any future development would, in fact, take the form of replacement or improvement of the existing plant.

Case Study 44
Buying land for the long term

In 1966, a pharmaceutical company based in Liverpool, acquired 10 hectares of agricultural land in Wirral for the construction of a new factory to which it relocated its Merseyside operations. Although the site was far larger than necessary at the time, the company took a long term view, believing that the land would meet its expansion requirements for many years. This view has proved correct. The company is now part of an international conglomerate and the site has been developed progressively over the years. A series of planning applications, involving the alteration or expansion of the premises, were submitted between 1985 and 1989, the most significant of which proposed a 1200 square metre extension to the main production building. Construction commenced in November 1989, shortly after the approval of the planning application, and the project was expected to be completed in 1991. Approximately half the site, however, remained undeveloped. Although the company still had plenty of space to expand, future investment in Wirral would be determined by the parent company's international strategy.

long term land holdings. Of the 23 sample sites developed by major manufacturing interests, 74% were owned by the company four years prior to development. The comparative figure for minor manufacturing interests was 26%.

The locational decisions of most of the major manufacturing interests interviewed reflected a long term perspective of their future land requirements and sufficient land was acquired to enable expansion (Case Studies 43 and 44). However, other research (Adams, 1990) has suggested that smaller companies take only sufficient space for their immediate needs. When product demand increases, company growth may well be constrained unless the firm is able to acquire adjacent land or to move. Evidence suggests that successful firms may move several times as they climb up the property ladder and examples of this have been given elsewhere (Case Study 60). Normally, the area of search on relocation is restricted to within two or three miles of the existing operation, almost two thirds of all the manufacturing companies interviewed considering ties to their existing workforce to be a critical factor at their current location. Indeed one company felt that it was so important to retain its existing work force that it would be loathe to relocate from its existing site, even though the only vacant land within the site was so heavily constrained that it made further expansionary development uneconomic (Case Study 45).

Case Study 45
Company's ties to local workforce

This 3.2 hectare site on an established industrial estate in Crewe was first developed in the 1950s. It was acquired freehold by the present owners in 1978, with the benefit of an existing factory unit and machinery.

The site afforded sufficient space for further development and in 1986 a major extension to the factory and plant room was built, providing 1700 square metres of new floorspace. Although approximately 0.8 hectares still remained, further development was severely constrained by poor ground conditions. This land was poorly drained and often waterlogged. It could not support heavy structures without substantial additional costs. The company, however, had no immediate plans to expand but if it did, it would not wish to relocate outside the borough, because of its existing workforce.

The particular requirements of an individual company, and the constraints and costs these impose on relocation, may in some cases lead even minor manufacturers to acquire and retain land for future expansion. The cost and feasibility of relocating existing operations will be an important determinant in companies' land management strategies. A long established marine and chemical engineering company in Wirral, for example, acquired the site of the adjacent concrete batching plant when it became available in 1986, although it then had no plans for its immediate development. The company was thereby able to expand its production facility when required two years later. In this case the relocation of the company would have been heavily constrained by its dependence on a dry dock which was located one mile away from its main site (Case Study 91).

It is important to briefly acknowledge this dimension of the heterogeneity of the manufacturing sector. Companies' behaviour in land and property markets is determined not only by their size and corporate structure but also by specific production or process requirements. Some companies may be constrained to acquire sites within very closely defined submarkets — the manufacturer of soft drinks for whom a spring of fresh water was an essential factor; the chicken reprocessing plant which was constrained in its choice of site by the relatively unpleasant nature of its activity; and the manufacturer of dairy products for whom available milk allowances and a suitable production facility were essential ingredients (Case Study 46). The way in which such specific requirements affect transactions will be discussed in Chapter 9.

Case Study 46
Site-specific production requirements

This 1.4 hectare site was in a rural settlement near Chester. It was first developed for the manufacture of dairy products in 1916, but changed hands several times as a result of company takeovers. In 1987, the site was bought by the largest independent manufacturer of processed dairy products in the UK.

This acquisition represented a strategic expansion of the latter company's operation. Expansion by any manufacturer of dairy products is constrained by the allocation of milk yields by the Milk Marketing Board. The takeover of an existing operation removed the necessity to have milk re-allocated from another producer. The purchaser was...

...further attracted to this particular plant by its reputation for quality and its established trade links.

Much of the original factory was subsequently demolished and was replaced by a prestigious new building, opened in 1990.

One implication of substantial landholding, by major manufacturing interests in particular, is that large areas of potential employment land may remain under utilized for many years. Whether such extensive landholdings are developed, released or simply remain idle clearly depends to some extent on structural influences, such as the demand cycle. As production requirements change and new products are developed, plans for utilizing expansion land may ebb and flow. Furthermore, technological change will also have an impact on companies' land requirements. Although two significant developments had recently taken place at the Vauxhall car plant at Ellesmere Port, for example, a substantial area of undeveloped land was thus declared surplus in the early 1990s, having been held as future expansion land for almost 30 years (Case Studies 47 and 48).

Case Studies 47 and 48
Expansion land surplus to requirements

The long term view which major manufacturing companies have of their future land requirements can take little account of the impact of future changes in technology or production processes, or in the case of multinational companies, of the comparative advantages of locating parts of the production process in other countries. A later re-appraisal of production requirements and development plans may release land which has been idle for many years.

In 1961, Vauxhall Motors acquired a former RAF airfield of 160 hectares on the boundary between Ellesmere Port and Wirral and began to build the company's new Merseyside car plant. At that time, Vauxhall intended that it would eventually use the entire site, but less than half was needed for the initial development, completed by 1964. By the early 1990s, the operational area of the car plant still covered only 80 hectares. The two most significant recent developments had been the construction of a new corrosive plant of over 2500 square...

...metres on a greenfield plot in 1986 (Site 47) and the redevelopment of an existing paint shop (Site 48), completed in 1990.

By 1990, it had become apparent to Vauxhall that the global changes which had taken place in vehicle manufacturing processes since the 1960s meant that all future investment plans could be accommodated through increased productivity within the 80 hectare operational area. The company was therefore able to declare the other 80 hectares, beyond the operational boundary, surplus to requirements. This land fell within the administrative area of Wirral MBC and would provide the largest single site suitable for industrial development to become available in Wirral for over 20 years. The local authority considered it to be of strategic significance within the Merseyside and north Cheshire context.

Detailed discussions were therefore undertaken between Vauxhall and Wirral MBC in the early 1990s and 35 hectares were identified for industrial development in the unitary development plan. This was intended to be reserved for major employers requiring sites of a size which could not be readily accommodated elsewhere in the borough.

The preponderance of multiplant and multinational companies within the sample showed that the redeployment of production at a regional, national or international level often determined development decisions at the local level. Such companies were often reluctant to dispose of unused expansion sites, for fear that to do so would jeopardize substantial existing investment in the plant, by constraining future development. In such cases, exchange values are exceeded when the implied cost of risking future flexibility is added to existing use values. National and multinational companies with a substantial portfolio of landholdings in many locations, thus tended to ignore the market potential of unused but relatively small areas alongside existing factories. Whether such behaviour can be said to be rational or irrational can only be judged against the uncertainty often surrounding future company plans.

A number of catalysts, however, did encourage manufacturing interests to consider releasing land. One factor was obviously the re-appraisal of land requirements in the light of technological change as already discussed. Another was the eventual acceptance of the permanent nature of the industrial restructuring of recent years, as in the case of Cammell Laird's decision to release 9 hectares of surplus land in the late 1980s, prior to the eventual closure of the company (Case Studies 106 and 107). Vacant land was also released following corporate restructuring (Case Studies 49 and 50) or, in the case of certain former nationalized industries, as a result of privatization.

Case Studies 49 and 50
Corporate restructuring and the release of surplus land

Corporate restructuring is one factor which may release surplus expansion land. In this example, 28 hectares may thus become available in an area where the local authority had identified a shortage of much needed industrial land.

In the 1950s, a well known food manufacturer acquired 46 hectares of land in north Wirral. The new factory constructed here enabled the company's output, previously concentrated in the Midlands, to be expanded substantially. By the 1980s, the operational area of the factory extended to 18 hectares (Site 49). Outside the operational area, the remaining 28 hectares of undeveloped land is divided into two parcels: 7 hectares to the east (Site 50a) and 21 hectares to the north (Site 50b). In 1986, the local management at north Wirral bought out the operation from the parent company, as a result of which brand names were subsequently produced under franchise. As part of the buy out, the new company formed by the local management acquired a ten year lease on the entire site and buildings, at an annual rent of £500 000, with an option to acquire the freehold in 1996 at a fixed price of £4.3 million.

The plant continued to prosper. Indeed a new warehouse of 1300 square metres was added in 1989 on the site of former railway sidings within the operational area (Site 49). At about this time, however, the company came to the view that future production requirements could be satisfied within the existing operational area, thereby allowing the 28 hectares of undeveloped land to be released. This decision was reached despite the fact that almost the entire operational area was now in productive use. It was hoped that the future sale of the 28 hectares would raise sufficient capital to cover the cost of freehold acquisition.

Before any marketing could take place, complex site constraints needed to be investigated, and if possible resolved. The company also entered into lengthy discussions with the local planning authority and commissioned consultants to study alternative uses for the 28 hectares...

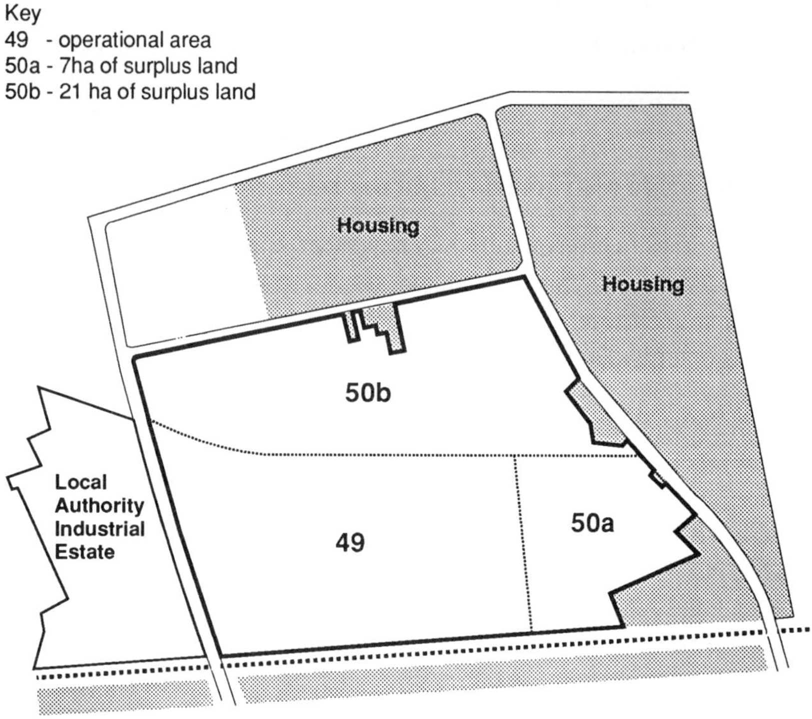

Fig. 6.1 Surplus expansion land

...The local authority was anxious to achieve an overall strategy for the whole of the 28 hectares, which would avoid piecemeal development and relate new uses sensitively to the surrounding area. Almost all the developable land on a popular nearby industrial estate had now been taken up. According to the local authority, there was a serious shortage of available industrial land here in the north Wirral and the surplus 28 hectares offered the only real chance to satisfy the demand for well located industrial land in a pleasant environment outside the green belt.

The local authority was keen to see the supply of available industrial land in this part of Wirral increased. However, without the corporate restructuring which took place in 1986, and the opportunity which it presented for the local company management to acquire the freehold, the release of surplus land may not have been considered within this timescale.

More importantly, the development boom of the late 1980s increased the incentive to sell as land values rose in response to increased development pressure. In such cases, companies which had previously concentrated solely on the use value of potential expansion land began to be influenced by rising exchange values. This had two important effects. First of all, some production interests endeavoured to market their surplus land for higher value uses, such as retail development (Case Study 14). It has been shown how this behaviour can undermine the potential supply of industrial land. Other companies responded to trends in the property market by themselves being involved in speculative development and in these cases behaved primarily as a financial interest. This particularly applied to 'other production interests', most notably those concerned with declining forms of water transport discussed below, but examples have also been given of manufacturing interests engaging in speculative activity (Case Studies 40 and 57).

In summary the ownership behaviour of manufacturing interests is determined by a variety of factors — from wider structural change to changing individual corporate objectives; from cyclical economic change and technological innovation to the specific requirements, constraints and optimizing production strategy of individual companies. Company landholdings are influenced by projected long term growth balanced by the financial significance of land as an asset in the company's portfolio, and ultimately by a financial appraisal of the advantages or otherwise of holding land, or indeed of speculative behaviour in the land and property markets.

6.3.2 Other production interests

'Other production interests' were dominated in the sample by those companies concerned with declining forms of water transport — the Mersey Docks & Harbour Company (MDHC) and the Manchester Ship Canal Company (MSCC). These companies have substantial landholdings in Wirral and Ellesmere Port respectively and were usually represented in the sample through their passive ownership of potential development sites.

Although the operational land requirements of this sector had declined substantially as a result of wider structural processes, their own corporate arrangements and objectives often militated against rapid land release. For instance, the proceeds of any freehold sales by the Mersey Docks & Harbour Company previously had to be returned to shareholders and could not be re-invested in port modernization. Whenever possible, disposals by the company were therefore restricted to a leasehold basis. This policy had been established for operational as well as financial reasons, since the company sought to retain freehold control of land for the possible future expansion of the port. In fact, when the National Dock Labour Scheme was abolished, the company reported greater interest in port related development. It was therefore less willing to release its under utilized land for nonport related activity (Case Studies 51, 52 and 53), but there are also important examples of diversification by this company in response to trends in property markets (Case Studies 54 and 112).

Case Studies 51, 52 and 53
Retention of land for operational use

To achieve their economic objectives, local authorities are concerned to ensure an adequate supply of available land. In Wirral, the borough council is keen to encourage the Mersey Docks & Harbour Company (MDHC) to rationalize its substantial landholdings within the docks and release sites for other employment development. In contrast, MDHC whose income is almost entirely port related, has seen little point in releasing land other than to companies which contribute to an expansion of port activity.

These three sites, for example, of 2.4 hectares, 0.2 hectares and 0.6 hectares respectively are located within Wirral docks. They have each been identified by Wirral MBC as suitable for industrial development but are being retained by MDHC for port related uses.

Most of Site 51 had in fact been leased since 1985 to a manufacturer of animal feeds for storage purposes and one small building on part of the site was occupied by an engineering firm. Existing Victorian sheds covered only perhaps a third of the area but in reality it had no further development potential while used for port related storage.

A number of approaches have been made to MDHC to lease Site 52 but without success. At one time, MDHC were prepared to consider the possibility of developing workshop units on Site 52 which is adjacent to the mill conversion discussed in Case Study 54. When the property market later declined and port activity became more buoyant, the company decided to retain the land for operational purposes as it had a usable berth.

For most of the 20th century, a flour mill with ancillary warehousing, occupied Site 53. It was operated by a food processing...

...company, under a 99 year ground lease from MDHC. When the buildings became obsolete in the early 1980s, MDHC agreed to the surrender of the lease, provided that the food processing company was responsible for demolition. By mid 1985 the site had been cleared, fenced and handed back to MDHC. MDHC maintains a pool of sites for temporary storage and cargo movement and, since 1985, Site 53 had been retained for these purposes, given its proximity to a berth.

Case Study 54
Speculative venture by production interest

Buoyant land and property markets may encourage production interests to reappraise their landholdings and encourage them either to release land or even to undertake speculative development themselves. The refurbishment and subdivision of this large redundant warehouse into small workshops represented an unusual step for its owner, Mersey Docks & Harbour Company (MDHC). Indeed, its reluctance to release land for other than port related activities is widely known. In 1984/1985, however, MDHC was considering opportunities to diversify its business and both the strong demand for small workshops at that time and the availability of a vacant single storey warehouse of approximately 2700 square metres tempted the company to 'put its toes in the water' of property development.

Planning permission was granted in October 1985, but commencement of work was delayed for a further ten months while negotiations took place with the Department of the Environment (DoE) to secure an urban development grant. Differences between MDHC and the DoE over the treatment of profits within the calculations were eventually resolved and an urban development grant was awarded which covered 35% of the £600 000 costs of refurbishment and conversion. The development would not then have been economically viable without the grant.

The conversion demanded considerable new steelwork inside the existing frame and the recladding of the east and west elevations. The first tenant took possession in January 1987 prior to the final completion of the project, but because of a lull in the market during 1987, most of the units were not let until 1988.

Where they have considered releasing once operational sites, both MDHC and MSCC have sought to share in the development gain by participating in the development process rather than simply disposing of the freehold. MSCC in particular has diversified into property development. One effect of this has been that MSCC has attached a relatively low priority to the marketing and development of its land in Ellesmere Port because of the higher rates of return on sites in Manchester, particularly for office development (Case Study 108). This wider perspective and duality of interest on the part of production interests can thus be seen to inhibit the release of sites whose development potential for operational use might well be appraised more favourably by manufacturing interests.

6.4 FINANCIAL LANDOWNERS

The research discovered only limited involvement by financial landowners in the Cheshire—Wirral corridor, whether by financial institutions or private development companies.

6.4.1 Financial institutions

Financial institutions were almost completely absent. Most institutional investors are unwilling to risk their funds in industrial development in the north-west. In Greater Manchester, Roger Tym & Partners (1989) attributed poor investor confidence to the lack of comprehensive, well presented and up-to-date information, in the absence of which, investors either under valued opportunities or were unaware of them. Such a database would, however, need to provide investors with more detailed information than mere aggregate statistics. If too much attention is given to the quantity rather than the quality of available industrial land, institutions may be presented with the mistaken impression that a substantial over supply exists. This would offer institutions little prospect of continuous rental growth and would further deter their involvement. In reality, good quality land suitable for the prime development is difficult to find.

Only one sample site was therefore owned by a major institution and one by a minor investor. The former site had been in this ownership since 1981 and had still not been fully developed by the end of the decade (Case Study 55). Although similar examples are few, it does demonstrate how the land management and development strategies of financial interests, and their speculation in the land market, may lead to potential industrial land in a prime location remaining undeveloped for many years, despite increasing market demand in the late 1980s.

Case Study 55
Strategy of a financial institution

In 1981, a financial institution acquired a 3 hectare site in Crewe, on 1.8 hectares of which it developed six industrial units to let. In 1990, the remaining 1.2 hectares was still undeveloped. In the interim period, the financial institution had attempted to maximize its return in several ways.

Planning permission was first sought on two occasions for nonfood retail development, but was refused and the decision upheld at appeal. In 1989, outline planning permission was given for a petrol sales forecourt, car showroom and workshop unit on that part of the vacant land fronting the highway, together with 4500 square metres of industrial units to the rear. The inclusion of petrol sales and a garage had the effect of increasing the site's potential value. If marketed separately, the owners' agent would have expected to achieve £1 million for that part of the site alone, an area comprising perhaps 0.5 hectares.

The whole site of 3 hectares however, was marketed in 1989, with offers in excess of £3 million sought for the freehold. The decision by the financial institution to dispose of the entire site, despite the leases on the tenanted units being highly reversionary, apparently reflected its view that, in comparison with other areas nationally, Crewe was not a high growth area.

6.4.2 Private development interests

Although some prestige developments in the Cheshire—Wirral corridor were promoted by regionally based development companies, there was almost as little interest in industrial development in the area from the country's top development companies as there was from its financial institutions. As discussed and illustrated in Chapter 4, the most active development interests in the corridor were small scale local entrepreneurs.

They operated in a very different sector of the market from major companies and institutions. Their local knowledge enabled them to move fast, spot opportunities and take advantage of the rising property market. They were prepared to acquire, refurbish and subdivide older buildings and undertake development on problematic sites. Refurbishment on one part of a site was often followed by a new development on another part.

Such local developers, like all development interests, acted in response to rising rents and the expectation of future rental growth. However, by approaching market analysis from the local level, they were able to take a more optimistic view of development prospects than those financial and development interests based outside the region. The findings in the Cheshire—Wirral corridor thus confirmed other work (Price Waterhouse, Casco and Chesterton, 1989) which identified the importance of local market knowledge in getting development schemes started in inner urban areas. Indeed, such local developers appeared to have the necessary flexibility to react quickly to changing market circumstances (Case Study 56) and their role in the market has been primarily as purchasers of land rather than as vendors. Such small local companies will tend not to hold substantial reserves of land for future development. However, it will be inevitable that some proposals may be shelved in response to economic changes, leaving undeveloped land within company portfolios (Case Studies 57 and 58).

Case Studies 56 and 57
Market appraisal by local interests

In 1987 a site of 1.4 hectares in Wirral was acquired by a local development company (Site 56). The site comprised an existing disused factory unit with 3000 square metres of available floorspace, together with an additional 0.8 hectares of vacant land (Site 57). The factory had been empty for two years but, although vandalized, it was structurally sound. It was successfully refurbished and subdivided into six units which were completed in 1989 and opened by the Mayor in 1990. All units had been successfully let. This was the first such project by this small company and satisfied an identified gap in the market at a time of good demand.

In contrast, the surplus land (Site 57) was sold freehold by the developer to a manufacturing company which had taken one of the refurbished units. Their intention was not to retain it for their own...

Fig. 6.2 Successful conversion and refurbishment by local development company in Wirral (Case Study 56)

...expansion, but to develop it for office, light industrial or warehouse use, for sale or lease. It had been hoped that development would have been undertaken in 1990. This has been delayed indefinitely by the general economic climate.

Case Study 58
Local developer responds to market conditions

In November 1988 a leather and skin company, which had relocated within Chester, put the site and building which it had previously occupied on the market. The existing building and the site of 0.4 hectares on which it stood were held on a 125 year ground lease from 1967. The lease was offered for sale at £220 000.

An offer of £190 000 was received from a local development company, who wished to refurbish, subdivide and let the premises. The lower figure reflected the poor visual appearance of an otherwise structurally sound building, the cost of the proposed work and the prospective rents which the development company believed could...

...be obtained. After some negotiation, a price of £208 000 was agreed and the site and buildings were acquired by the development company in May 1989.

In November 1989, the company obtained outline permission to subdivide the main building into two and construct one new industrial unit within the yard. One half of the main building was prelet before work commenced in late 1989. After six months of marketing, the other half of the building was taken by an engineering company in the spring of 1990 on a 25 year lease. By this time, the property market had slowed down, however, and the developer decided not to proceed with the construction of the new unit, but rather to make the land available to an adjacent firm for storage, thereby generating some rental income.

6.5 STATUTORY LANDOWNERS

Statutory landowners were at the forefront of reclaiming, servicing and making industrial land available to production interests. Local authorities were the most significant statutory landowner in this respect. The role of local authorities in the provision of industrial land is related to their wider responsibilities for economic development. All four district authorities sought both to attract inward investment and to encourage the formation and expansion of local businesses. English Estates and the Merseyside Development Corporation were important but were involved to a more limited extent, at least in terms of the number of sites involved if not the overall land area. The discussion below will focus on the activities of the local authorities and these statutory development agencies. Other statutory landowners were relatively insignificant. One undeveloped site was owned by the British Rail Property Board. Two further sites previously in this ownership category had in fact been released, one by British Rail and one site, now redeveloped, had previously been owned by an area health authority.

6.5.1 Local authorities

Statutory landowners make an important contribution to the process of industrial development. They have a strategic role in assembling large sites,

Fig. 6.3 Crewe Gates Industrial Estate

Fig.6.4 Tarran Industrial Estate, Wirral

both to meet current user demand and to guarantee long term supply. In Crewe, for example, local authority industrial estates have a long history. In the late 1960s, the borough council opened up the Weston Road area for this purpose. In the mid to late 1970s it acquired and laid out the Crewe Gates Industrial Estate. In the late 1980s, it entered into partnership with the county council to promote the Crewe Business Park on land owned by the two authorities. Similarly, Wirral MBC's role in making land available at Tarran Industrial Estate has been important over many years in enabling the relocation and in situ expansion of smaller local companies (Case Studies 59, 60 and 61).

Case Studies 59, 60 and 61
Meeting the needs of minor manufacturing companies

These three case studies relate to separate plots within a large industrial estate. The land was owned by the local authority and it was in a minor urban area of Wirral. It was designated for employment use in 1967, since when it had been serviced progressively. In 1990 it was almost fully developed. The estate has proven very popular with local businesses seeking to relocate to better premises and half the total area was developed to provide for the in situ expansion of existing companies. These points are demonstrated by the three examples below.

The first is a plot of 0.36 hectares, the freehold of which was acquired in 1983 by a manufacturer of security systems. Part was immediately developed for its own use but the majority remained vacant until 1985/6. Four further industrial units were then built, providing an additional 2500 square metres of floorspace. At that time it was intended that only two units would be for the company's own use. The others would be let. Thus it can be seen that manufacturers may also acquire and manage land as part of their portfolio of assets rather than to meet their own production requirements. In 1990, however, all units were occupied by the company. The plot was then fully developed.

The second plot, 0.22 hectares in area, was on one of the later phases of the estate. Being reclaimed land close to a river, it suffered from poor ground conditions. These could be overcome with piled foundations but affected the marketability of this and nearby plots. Indeed, in this case, development was only feasible with financial...

> ...assistance from the local authority, covering 50% of the costs of foundation works. Nevertheless, a factory development of 420 square metres made possible the expansion of a small local company which started in a back room ten years previously and then relocated to a smaller unit on an earlier phase of this same estate.
>
> The third plot comprises 0.41 hectares acquired on a 99 year ground lease in 1987. The completion in 1988 of a 750 square metre factory facilitated the expansion of a small manufacturing company established on the estate, 15 years previously. It had extended as far as possible within its existing site. Although not adjacent to the latter, this nearby plot met its additional requirements.

One of the most important large sites in the Cheshire—Wirral corridor brought forward by the public sector in the late 1980s was the 34 hectare Chester West Employment Park. Development was considered by the local authority as early as 1976 but was delayed for some years by the cost of tackling problematic physical constraints. In particular, the site was poorly drained, with a high water table which restricted the depth of foundations. Chester City Council subsequently spent a total of £4.5 million on reclaiming, servicing and laying out the first phase of the park, including the provision of a new pumping station. Much of this expenditure was supported by finance from the European Regional Development Fund.

It will be evident from the many case studies which involve statutory landowners, and from the later analysis of transactions in Chapter 9, that the public sector, far from being a passive landowner, plays an active part in development activity itself and in making land available to the private sector. There are, nevertheless, supply constraints on some sites marketed by the public sector, insofar as they are available on terms and conditions which are restrictive or which prospective developers may find unfavourable. The expression of the local authority's economic development objectives in its marketing strategy, for example, can result in land vacancy and low rates of take up. This was evident in Crewe (Case Study 62) and in Chester, where the city council has been particularly keen to exercise close control over the type and quality of users allowed on Chester West Employment Park. It was therefore only prepared to dispose of land on 999 year ground leases, in order to retain management control. Although speculative developers were not opposed in principle to leasehold acquisition, the length of the draft leases (each of 40—50 pages) and their restrictive nature (prohibiting distributive uses and certain other activities) caused considerable concern and resulted in substantial delays to some schemes (Case Studies 63 and 64).

Case Study 62
Local authority marketing strategy: patience with manufacturing company

This site of 0.7 hectares was located on one of the later phases of the Crewe Gates Industrial Estate to be brought forward and marketed by the local authority. Although it had been under option to a local manufacturer of polymers since 1987, and planning permission for a new factory of 3000 square metres was granted in 1989, acquisition of the ground lease was not completed until 1990. A number of factors contributed to this delay.

First, the prospective purchaser felt no urgency to acquire the site. The company occupied four other units locally and relocation to the new site would simply remove any problems caused by a split operation and provide sufficient land for future expansion. Secondly, the company had experienced financial difficulties in the immediate past, with a loss of orders having led to redundancies. It secured financial assistance from Crewe Development Agency, in the form of an equity injection of £100 000.This would be paid upon receipt of invoices during the construction period, but was administered in such a way as to improve the book value of the company's assets. Prior to this, the company was at the maximum gearing which the bank would allow. Thirdly, development depended critically on the approval of derelict land grant to treat puddling clay on site. A decision on this application was anticipated in September 1990. Finally, disputes had occurred with the architects during the design stage.

Some local property agents criticized the local authority practice of allowing companies to 'option up' sites at no cost for considerable lengths of time. The local authority, however, places great emphasis upon enabling existing local companies to expand and remain in the area. Furthermore, the relocation of this company would release three units which it rents from the local authority which could then be relet.

Case Studies 63 and 64
Local authority retains control though ground leases

Although several of plots on Chester West Employment Park had been developed by 1990, these case studies focus on two undeveloped plots of 2.4 hectares and 1.8 hectares respectively. In 1989, planning permission was given for 4000 square metres of B1 development on the larger of the two plots (Site 63) and for 7500 square metres of industrial and office space on the smaller (Site 64). Both were speculative proposals by private development companies. The larger scheme was intended to be built in phases in response to market demand and would provide more traditional industrial accommodation than elsewhere on the park.

Although both developers first expressed interest in the sites in 1988, neither had completed acquisition by mid 1990. A particular stumbling block during the intervening period proved to be the terms and conditions proposed to be attached to the 999 year ground lease, which both the developers and the local agents considered to be onerous, particularly in precluding warehouse development throughout the site. In 1990, however, the city council agreed that warehousing could be permitted for a single user, of up to 1000 square metres of floorspace. Nevertheless, by this time, the downturn in the economy and high interest rates led to increased hesitancy about undertaking speculative development and there appeared little likelihood that either scheme would be undertaken in the short term.

Conversely, the terms of a lease can effectively inhibit private land speculation and ensure the development of the site (Case Study 65). Indeed, local authorities within the study area sold land subject to conditions in the lease which ensured rapid development not only by private interests but also by other statutory developers such as English Estates (Case Study 66).

As well as assembling land, supplying development sites and providing infrastructure, the public sector has a second and equally important responsibility. This involves acquiring land left derelict, often by former industrial users, recycling it and making it available for new development. There is much potential for such activity in the corridor, especially in

Ellesmere Port and Wirral. Ellesmere Port & Neston Borough Council has achieved a particularly successful record in acquiring and reclaiming derelict industrial land (Case Study 85). Such work is dependent, however, upon the provision of derelict land grant by central government. Apart from sites acquired with derelict land grant, Ellesmere Port & Neston Borough Council, for example, owned few potential industrial sites of its own.

Case Study 65
Local authority marketing strategy: preference for manufacturing company

Crewe & Nantwich Borough Council has played a vital role in recent years in bringing land forward for industrial development. This is reflected in the many new and more established companies which have been able to obtain land and property from the local authority to meet their specific needs. Existing companies, in particular, have been enabled to acquire adjacent land or relocate within estates to meet their changing requirements. The particular economic objectives of the local authority were reflected in their marketing strategy, and in the terms of land disposal. Sites were normally disposed of initially on a leasehold basis, conditional upon the site being developed within a specified period. When the development was complete, the freehold would be available but in the event of development not taking place, the local authority retained the right to repossess the site at the sum originally paid for it.

This case study concerns land on the Crewe Gates Estate, leased in 1990. The lessee subsequently indicated that he no longer proposed to develop the site but wished to assign the lease to a developer to construct small units. The local authority was unwilling to approve this since there was a queue of small manufacturing companies seeking plots of this size for their own use. Indeed, one such company, whose application was likely to be considered favourably by the local authority, occupied a smaller plot on this estate but needed to relocate in order to expand and develop a new product. The strategy of the local authority reflected its commitment to facilitating the expansion of established local companies and ensuring the retention of existing jobs within the borough.

Case Study 66
Local authority development strategy reflected in restricted terms of disposal

In 1976, Ellesmere Port & Neston Borough Council acquired 7 hectares of land, part of which provided an extension to the existing Clayhill Industrial Estate at Neston. The local authority itself built units on part of the site, and once reclamation and servicing were completed, offered the remainder for sale.

English Estates expressed interest in the acquisition of 1.7 hectares in order to construct small workshops. The local authority was only prepared to sell land with a strict timetable for development, in order to prevent speculation by the private sector. It was not prepared to waive this condition for English Estates. To overcome this, English Estates acquired an initial 0.6 hectares in 1986, on which 13 workshops were completed by the end of 1987 and quickly let. A further hectare was then acquired in 1988, on which seven larger units were completed by September 1989, with a further two under construction in 1990.

Local authorities in the Cheshire—Wirral corridor, particularly in Wirral and in Ellesmere Port & Neston, have also been active, wherever possible, in developing their own land, notably to provide smaller starter and nursery units identified as being in short supply (Case Study 67). Although it undertakes careful market assessment and is unwilling to build without good evidence that units will be occupied, the public sector is less concerned about future rental growth than private development interests. Sites in public ownership have therefore frequently been brought forward which would not have been an attractive market proposition for private development interests — sites either in need of reclamation work or in areas where investment confidence was weak (Case Study 68).

It will be evident from many of the case studies, however, that the land reclamation and development programmes of local authorities depend critically on the availability of public finance and particularly, with increasing constraints on local authorities' own capital expenditure, on central government and European funding. Indeed, several case studies show how financial constraints have necessitated the phasing of relatively small developments over many years, how later phases have been either delayed (Case Study 6) or abandoned by the local authority (Case Study 68) and how economic development objectives can be undermined by resource constraints, with potential employment sites in danger of being sold for higher value uses (Case Study 99).

Case Study 67
Local authority develops small units

Fig. 6.5 Conversion and new units built by local authority (Case Study 67)

The behaviour of the local authority shown in this case study, in actively developing its surplus land for industrial use, can be contrasted with the relative inertia of many large manufacturing companies with regard to shedding nonoperational land.

Site 67 was a former local authority depot, vacated when council premises were rationalized. Development of the site to provide starter units was undertaken in four phases by Ellesmere Port & Neston Borough Council, with work commencing in 1983. With the exception of some small sheds which were demolished, existing depot buildings were retained. Early phases involved the 'creative conversion' of three sided sheds. Phase 3, completed in 1988, comprised the conversion...

...of an existing building to form ten small units, providing over 1800 square metres of floorspace. Phase 4, however, was entirely newly built, providing a further 460 square metres in three small units. In 1989 the site was fully developed and units had been let successfully.

Development was facilitated by considerable grant aid. An ERDF grant of £135 000 was given towards infrastructure costs and urban programme grants totalling £600 000 covered 75% of the development costs of Phases 1 — 3. The local authority was later no longer eligible for urban programme grants and therefore funded Phase 4 itself.

Case Study 68
Redevelopment of local authority site

Although this 2.75 hectare site was within a rural part of Chester district, it had been owned for many years by Ellesmere Port & Neston Borough Council. It was formerly a sewage treatment works, but this ceased operating over 20 years ago. In the 1970s the availability of better sites elsewhere deterred the development of this site which experienced poor ground conditions. After 1981, it came to be identified with the proposed reclamation by the local authority of a large nearby site, formerly owned by Burmah Castrol (Case Study 9). It was anticipated that this smaller site would be brought forward as part of a much larger project.

The reclamation of this site was brought forward, however, when in 1987 a contractor working in the vicinity was looking for somewhere to dump excavated clay. Permission was given for it to be tipped here, subject to satisfactory compacting and rolling. The site was then serviced with the benefit of a derelict land grant of £300 000 awarded in 1987. The local authority subsequently built five units of approximately 100 square metres each on an area of 0.17 hectares. Although it would have liked to build further units, in 1990 the remaining land was being marketed because of capital constraints and negotiations were underway with a local building company.

The important role of local authorities in one of their key economic development activities is being threatened by wider central government objectives and constraints on public expenditure. Ellesmere Port & Neston

Borough Council, for instance, which had carried on building small industrial units on its industrial estates throughout the 1980s, had to cease this activity in 1990/91, as a result of capital constraints.

In conclusion, much less industrial development would have taken place in the Cheshire—Wirral corridor in the years 1985 to 1989 without the involvement of local authorities. They have played an active role in the supply of land for industrial development, have enhanced the quality of potential sites through reclamation and road and infrastructure improvements and have attempted to anticipate and meet changing demands for industrial space through their development programmes. Broader political priorities will determine the extent of such participation in the future.

6.5.2 Statutory development agencies

(a) English Estates

Unlike private local and regional development companies, English Estates acquired and prepared sites, and maintained a land bank in anticipation of market demand. One of the concerns which underpins Wirral's unitary development plan is that, while much of the 1980s were characterized by serious recession, when there were signs of recovery later in the decade the supply of suitable land and buildings in some parts of the borough was inadequate to meet demand. Traditionally, English Estates played a key role in providing serviced sites and undertaking advance factory construction in the assisted areas. One of the most important such sites in the Cheshire—Wirral corridor is the Croft Industrial Estate and Business Park, comprising over 30 hectares, at Bromborough in south Wirral (Case Studies 69 and 70).

Case Studies 69 and 70
Action by English Estates

Croft Industrial Estate provides a good illustration of the dominant role played by the public sector, and especially by English Estates, in the provision of new industrial floorspace in areas which are less attractive to the investment market. When, in 1977 and 1987, English Estates acquired land and brought it forward for development, there was very little confidence within the private sector in the viability of speculative industrial development in this area. English Estates...

...was able to provide advance units and large fully serviced sites directly to users and thus, in the absence of private sector involvement, satisfy perceived local shortages.

Croft Industrial Estate is located off the A41 in Wirral. Development has been promoted by English Estates progressively over a number of years, in response to market demand. Originally 17 hectares (Site 69) were acquired in 1977 from a major manufacturing company with substantial holdings of surplus land in this area. It was a greenfield site with services available at the boundary, good access and no adverse physical conditions. Since acquisition, a mixture of workshops and larger advance factory units have been developed. Between 1985 and 1989, for example, planning permission was granted for over 10 000 square metres of industrial floorspace. Some plots remained available for immediate development but by the mid 1980s, English Estates recognized that its future ability to respond to demand would be limited by a lack of space.

In 1987, the owner of the freehold wished to take back approximately 0.8 hectares of the site in order to provide access to an adjoining proposed retail development. English Estates saw this as an opportunity to convert its 125 year lease into a freehold and also to acquire additional adjacent land. This additional area of 15 hectares (Site 70) was regarded by English Estates as a strategic acquisition in terms of size and location. Approximately £5 million was invested by English Estates on the costs of acquisition, servicing, infrastructure and landscaping. By August 1989, the site was available for development, with marketing directed at business users as well as traditional manufacturing companies.

At the time of writing, the role of English Estates is changing. Much of its existing property is being sold off to the private sector. Approximately £40 million was raised in this way in 1989—90, with a further £60 million of sales programmed in 1990—1. In 1990, the Government proposed that English Estates should move away from undertaking advance factory construction itself and instead encourage private sector provision, through grant aid. More recently, it was announced that English Estates would be absorbed from 1994 in the new urban regeneration agency, now titled English Partnerships. The changed role of English Estates, together with resource

Fig. 6.6 English Estates site on the A41 in Wirral (Case Studies 69 and 70)

constraints which it is anticipated will severely limit local authorities' own building programmes, will place greater emphasis on the role of the private sector in anticipating, and responding to, cyclical and structural changes in demand.

(b) Merseyside Development Corporation

Merseyside Development Corporation (MDC) was established in 1981. Most of its activity has been undertaken on the Liverpool side of the Mersey, as its initially designated area included only a very small part of Wirral Docklands. In 1988, the designated area of the Merseyside Development Corporation was substantially extended, incorporating a much larger slice of Wirral Docklands. The development corporation has since been much more active on the Wirral side of the river, and has played a key role in reclaiming sites which would not be attractive to the private sector (Case Study 71), in assembling sites in multiple ownership (Case Study 114), in overcoming passive ownership (Case Study 18), in providing land for manufacturing development and in providing financial assistance, without which some schemes would not be viable (Case Study 54). All of these measures have been important in overcoming supply constraints and have augmented the effective supply of land for industrial development.

Case Study 71
The role of Merseyside Development Corporation

This 1.4 hectare site in Wirral's inner area is an example of a site in which the private sector was unlikely ever to have been interested. The presence of four obsolete dry docks made it unsuitable for development without considerable reclamation work. Formerly a ship repair yard, it had been unused for some years and was actively marketed by Cammell Laird. It was acquired by the MDC in 1983. Site conditions were made more difficult by poor access to the public highway, the presence of an old tunnel cutting, filled land and outdated services and infrastructure. Site reclamation began in 1985/86.

By summer 1990, however, only one of the docks had been filled. This was completed in 1988, at which time it was anticipated that reclamation work would be finished in 1988/89 and the site would be under development in 1990/91. Filling the remaining docks was delayed when one of them was listed (Grade 2). According to English Heritage it is of historical significance. There are, however, no viable schemes for its re-use and its location does not lend it to being a tourist attraction. Nevertheless MDC still proposed to complete its servicing programme by 1992 and then to make the land available for B1, B2 and B8 development, for which planning permission was given in 1989.

6.6 SUMMARY

This chapter has offered a working classification of land ownership and development interests, based on the detailed knowledge of 120 case studies. It has described the diversity of motives and behaviour of the various interests which play a part in making land available for industrial development. The case studies have illustrated how the strategies of the different interests may respond to cyclical, structural or corporate changes, and how these strategies may in turn affect the supply of land for industrial development. In particular, it has been evident that the land management and development strategies of the key ownership interests have been subject to very different influences and considerations.

Historic landed property interests in the sample tended to be passive. Their sites have moved forward towards development only slowly and in response to approaches from other development interests.

The activity of financial interests in the Cheshire—Wirral corridor has been constrained by the relative performance of industrial property as an investment. This is influenced not only by the economic cycle, but also by comparison with the return on other types of property, the returns in other regions and the returns on other forms of investment. As a result much of the activity identified in the sample has been by local or regional development companies which have tended to acquire land for immediate development, and there has been little involvement by financial institutions.

The land management and development strategies of production interests tend not to depend on these wider comparisons. Development activity is determined by the need to meet immediate or changing production requirements, and is often cyclical. It has also been clear, however, that major manufacturing companies take a long term view of landholding, which overrides short term cyclical considerations. Trends in land and property markets may later release under utilized land. Similarly, noncyclical factors such as technological change, corporate restructuring or the reappraisal of global production strategies may also unlock land which had previously been retained for many years. Minor manufacturing companies are less likely to have significant landholdings. However, in exceptional cases they may take a long term view of future needs, where production processes make relocation difficult or expensive. Their behaviour, and the scale of their requirements are such that they have less impact on local land supply than major companies.

The statutory sector, unlike the other three interests identified, is active in the land market in ways which do not synchronize to the same extent with the economic cycle. In order to fulfil long term economic development objectives, it is more likely to assemble sites and to develop them in anticipation of demand. The key constraint here appears to be, not the immediate economic viability of land assembly, reclamation and development, but the availability of public finance. The release of land also reflects wider objectives and is not dictated by market trends alone.

This chapter has therefore highlighted the importance of moving beyond simple site based models of the development process, on which so many earlier policy initiatives have been based. As Healey and Barrett (1990) argue, it is instead necessary to explore how the strategies, interests and actions of individual agents in the development process relate to the much broader processes which frame or structure their decision making.

The next three chapters give particular attention to the actions which owners take, or fail to take, to challenge planning, physical, ownership and valuation constraints in order to bring land forward for development. Such action is analysed both from the perspective of the particular owner and in the light of the broader processes which structure the development of industrial land. The final chapter attempts to discern how far the actions, interests and strategies of individual landowners are constrained by the resources to which they have access, the rules which they considered govern their behaviour and the ideas on which they draw, in developing their strategies.

7

PLANNING FLEXIBILITY AND PLANNING RESTRAINT

At the end of 1989, approximately 161 hectares in the Cheshire—Wirral corridor (or 23% of the potential supply of industrial land) already had either outline or full planning permission. Planning applications were under consideration at that time on a further 39 hectares (or 6% of the potential supply). Thus, in comparison with the 164 hectares developed for industrial purposes between 1985 and 1989, some 200 hectares already had, or were likely to receive, planning permission for industrial development at the beginning of 1990.

This chapter examines the impact of town and country planning upon the market availability of land for industrial development in the Cheshire—Wirral corridor. Through the detailed case studies, it explores the extent of planning flexibility and planning restraint; investigates the contribution of planning agreements to industrial development; considers the loss of industrial land to other uses; and identifies the scale of pressure for major new greenfield development.

Before looking at the evidence on planning flexibility and restraint, however, some initial comments are necessary. First of all, planning 'flexibility' in this context should not be taken to imply a deviation in development control away from local plan policies and national planning guidance. Instead it reflects a sensitive response to the diverse land requirements of industry within development plans, reinforced by development control decisions.

Secondly, the discussion of planning restraint will draw particularly on development control records of planning refusals. McNamara and Healey (1984) drew attention to the potential limitations on the use of development control records, describing them as *'end-state statements of negotiative processes which are often complex'* (p. 96). In other words, such data gives no indication of the number of proposals aborted as a result of informal discussions with development control officers prior to any formal planning application. A small number of planning refusals, therefore, does not of itself necessarily indicate planning flexibility. Nevertheless, the analysis of development control data, interpreted in the wider context of development plan policies, will be indicative of the local planning environment.

7.1 PLANNING FLEXIBILITY

Local plans were discussed in some detail in Chapter 3, where it was shown that throughout the Cheshire—Wirral corridor, employment land allocations have reflected, and responded to, the impact of economic decline and restructuring on local employment. Thus, for example:

> *Provision is made for land for employment to be made available between 1986 and 2001, in order to meet the need to provide a stock of immediately available land. This includes a flexibility allowance to ensure a range of size and types and to capitalize on development opportunities, and is designed to give scope for new allocations* (Cheshire County Council, 1993a, Explanation of Policy EMP1).

A flexible approach towards industrial development was therefore much more evident than restraint in strategic planning policy in the Cheshire—Wirral corridor. At district level, local plans also responded to specific employment needs. The adopted Nantwich Local Plan (1987—96), for example, allocated the first major new employment site there in many years. A 7.5 hectare greenfield site on the edge of open countryside was identified to meet local economic objectives:

- to provide local employment opportunities for the people of Nantwich;
- to provide areas for the expansion of existing business which might otherwise be forced to move away from Nantwich;
- to allow for new employment development;
- to increase the range of employment sites available in the borough generally.

Because of the site's location, the development brief included provision for effective screening and landscaping (Fig. 7.1). Nevertheless, this allocation was essential to local economic development. Planning flexibility was also illustrated by several case studies in rural areas. On one site, seven small industrial units were constructed by a local manufacturing company on the

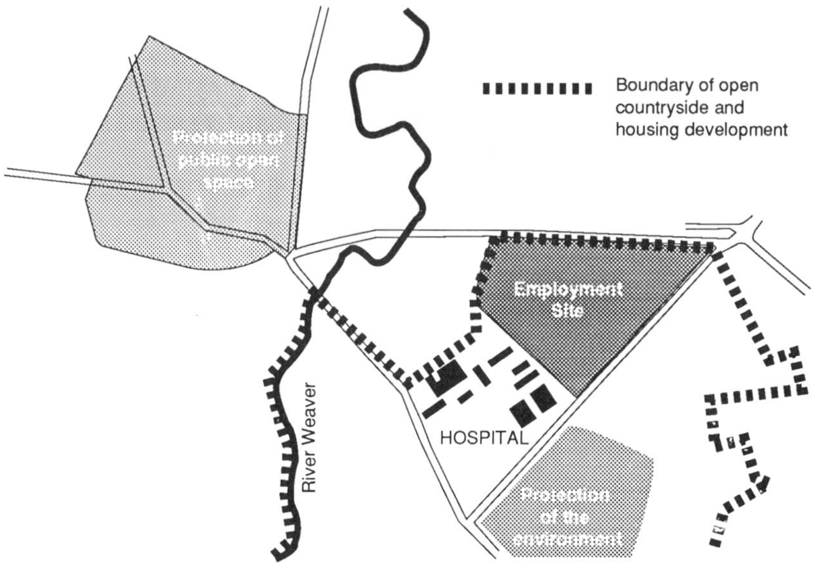

Fig. 7.1 Site allocated for employment development, Nantwich (Case Study 113)

Fig. 7.2 Conversion of redundant hospital in green belt to industrial units (Case Study 4)

edge of a village near Chester. The site had been in the green belt up to 1985, when it was identified in the Chester Rural Areas Local Plan as an inset in the green belt (Case Study 72). Changes in policy at both the national and local levels also permitted the change of use of redundant hospitals in the green belt (Case Study 4) and allowed a series of redundant farm buildings in the Cheshire—Wirral corridor to be converted to enterprise workshops, in order to enable rural diversification (Case Studies 73 and 74).

Case Study 72
Inset in the green belt

This case study demonstrates planning flexibility in the green belt. The allocation of the 0.4 hectare site in the Chester Rural Areas Local Plan 1985, as a light industrial employment inset in the green belt, made possible the development of 1500 square metres of new floorspace and facilitated the growth of the small, local manufacturing company which undertook the development. The company was also required by the local planning authority to construct five units to let, in addition to the two units which it intended to occupy itself.

The case study also shows how the development of sites identified by the local planning authority may be physically constrained even in rural areas. Difficult ground conditions, for example, added £3000 to development costs and necessitated piled foundations. Access presented a greater problem. The site was first approached by a B road, then by an unadopted road in poor condition, and finally by a private road owned by the water company which serves a nearby sewage treatment works. The company had to negotiate a legal right of way over the private road which added £11 000 to total costs and delayed construction by 12 months. Nevertheless the company, a manufacturer of electrical components, successfully relocated, expanded and retained its local workforce.

Case Studies 73 and 74
Planning flexibility allows rural diversification

Planning policy at both national and local levels has recognized the need for employment development and diversification in rural areas. These two cases demonstrate how it can be achieved by permitting...

...the redevelopment and change of use of redundant agricultural buildings.

Case Study 73 comprised the phased redevelopment of disused farm buildings and associated land on a site of 0.8 hectares within a working farm of some 57 hectares, to the south-east of Nantwich. Four planning applications were approved in 1987 and 1988, all of which had been implemented by the landowner by 1990. Three of these involved the refurbishment and conversion of existing farm buildings and one comprised new development to provide three rural enterprise workshops. The need for such units in rural areas was previously identified by the local authority and the development was encouraged by local authority officers. A total of 16 units were provided and successfully let for a variety of uses, such as cheese production, furniture restoration, car spraying and a removals business.

Case Study 74 relates to a 0.24 hectare site on part of a working farm which was divided by a road. It comprised a farmhouse together with a range of outbuildings too small for modern farming methods and in poor repair. Farmland and further buildings lay on the other side of the road. The site was on the edge of a village, south-west of Chester in the green belt. Policy E5 of the Chester Rural Areas Local Plan 1985 stated that *'There will be a presumption in favour of granting planning permission to use small vacant nonresidential buildings for light industrial or office purposes, subject to certain conditions'.* In 1990 planning permission was given to the landowner for the change of use of the farm buildings on the site to office and light industrial use. Whether the proposal would proceed, however, depended on the outcome not only of a structural survey and the costs of the necessary work, but also on whether planning permission would be given for the construction of a replacement farmhouse.

7.2 PLANNING RESTRAINT

Local planning policies and land allocations suggest that planning authorities have evolved strategies to both stimulate and enable employment development rather than hinder it, even in some cases in areas of traditional planning restraint. This conclusion is supported by an appraisal of development control data which shows that of the 182 industrial planning applications made between 1985 and 1989 on sites of 0.2 hectares or over, less than 5% were refused. Focussing on the 120 sample sites, there were only two examples where the refusal of planning permission delayed or prevented industrial development. There was therefore no significant evidence in the Cheshire—Wirral corridor to suggest that local planning authorities impeded industrial development.

Of the two sample cases where development was constrained by the planning process, one concerned not the sample site itself but a nearby plot of land (Case Studies 32 and 33). A complicated land swap had been arranged. One major manufacturing company, needing to expand, had arranged to purchase adjoining land from a neighbouring company. To do so, it had agreed to provide the neighbour with replacement land, on the other side of the neighbour's factory. This land, however, was used as open space. A planning application for industrial development was refused, following substantial public objection. This refusal was challenged by a second application which, after much effort, was approved and enabled the land exchange to take place. A key factor in the final decision was that, had planning permission not been given, a company which already employed 60 people and which planned to expand would have been compelled to relocate. The other case concerned an application by an existing company to relocate and expand within the green belt (Case Study 75). It was refused planning permission, not by the local planning authority, however, but after being called in by the Department of the Environment.

Case Study 75
Planning constraint on company relocation

In 1989, a planning application was submitted to Chester City Council for the development of 0.2 hectares of land in the green belt to provide an agricultural feed mill and storage building, together with a commercial vehicle repair workshop, providing a total of 750 square metres of floorspace. The site was part of an area of 3.2 hectares which had been owned by this family business since 1963 and on which there was a house used by the family. The proposal would have enabled the relocation of the business from existing premises in the conservation area of a nearby village, from which it had operated since the 1830s. The existing site included a large shed and a parking area for heavy vehicles, but access was difficult since the site was near the junction

...of two narrow cobbled village streets. The County Highways Department therefore welcomed the proposed relocation of the company to the new site (Fig. 5.1, p.114).

The proposed development was supported by the local planning authority, but was called in by the Department of the Environment as a departure from the development plan and an inappropriate development in the green belt. Following an inquiry, the application was refused permission. The proposal was not felt to represent an agricultural use and was considered likely to harm green belt objectives by setting a precedent in extending the built up area. The special circumstances and benefits of the proposal were not felt to outweigh that objection.

The company did not wish to relocate to an industrial site in either Chester or Ellesmere Port. It was particularly keen to retain its existing workforce and to minimize the transportation costs of a business which stored and dried grain and serviced commercial vehicles for the local area.

In certain limited circumstances, local highway authorities can direct a refusal of planning permission. The research discovered four instances where industrial development was delayed or prevented on highway grounds, rather than as a result of planning policy. Three of these sites were owned by former landed property. In each case, there appeared to be no urgency on the part of the landowner to challenge the highway constraint but as discussed in Chapter 6, this ownership interest tended to be passive and sites were being brought forward only slowly in response to development interest. One site was affected by a proposed new road while, in the other two cases, the proposed point of access from the site on to the public highway was considered unsafe by the authorities. In one of the latter cases, the owner had agreed to sell the site to a property development company. This company appealed against the refusal of planning permission on highway grounds and was successful. In the final instance another development interest, undertaking a scheme within Chester, had wished to construct a new access to improve the marketability of the site. The company received indications from the local planning authority, before any planning application was submitted, that the proposed access would be likely to be unacceptable on highway grounds. Amendments were thus made to the scheme at an early stage, keeping delays to a minimum.

7.3 PLANNING AGREEMENTS

During the period under investigation, planning agreements could be made under various legislation:

- Section 52 of the Town and Country Planning Act 1971; (now S106 of the Town and Country Planning Act 1990);
- Section 33 of the Local Government (Miscellaneous Provisions) Act 1982;
- Section 111 of the Local Government Act 1972;
- Section 278 of the Highway Act 1980.

Each power differed slightly in purpose and many planning agreements would refer to all four Acts. Agreements could be restrictive in their objectives, preventing certain uses of land or imposing restrictions on development. Alternatively they could be positive, requiring specific works to be undertaken. In either case, the purposes of the agreement would not normally be capable of being achieved through conditions attached to the planning permission.

Two examples illustrate how local planning authorities may attempt to gain greater control over the development process through the use of planning agreements than is possible through the exercise of development control powers alone. In one case, the relocation of a split manufacturing process onto a single and more suitable site was delayed while the local planning authority negotiated an agreement to ensure that the non-conforming use on one of the existing sites was extinguished (Case Study 76). In the other case, the local authority acted through a planning agreement to encourage industrial development to take place on a site which would otherwise have been left with no practicable means of access (Case Study 77). Planning agreements used in this way can clearly be an effective instrument to prevent the creation of vacant backland and the under utilization of a potential site. In this example, again the planning process can be seen not to have restricted the supply of employment land, but to have enhanced it.

Case Study 76
Acquisition delayed by Section 52 Agreement

This 2 hectare site which included an existing building with 1300 square metres of floorspace, was leased from Ellesmere Port & Neston Borough Council in 1989 by a printing company already operating elsewhere in Neston. The company had premises in Wirral, on which the lease was about to expire. Both operations were therefore relocated to this site, which enabled the company to overcome the problems of a split operation. The existing Neston operation however was a nonconforming use in a residential area. Acquisition of the new site took over a year to complete after the first formal expression of interest. Delay arose because of the negotiations over the Section 52 Agreement, which accompanied the planning permission in 1989 for change of use of Site 76. This agreement prohibited the development of the company's original Neston site for other than residential uses, in response to environmental objections arising from the proximity of that site to neighbouring houses.

Case Study 77
Development encouraged by Section 52 Agreement

Site 77 comprised 4 hectares, and was part of the 15 hectare site of a former steel mill in the inner area of Wirral. As a result of company takeovers and the rationalization of the steel industry, the mill itself was dismantled, sold and shipped overseas. The 4 hectares identified here had previously been occupied by storage bays and concrete slabs which covered perhaps 20% of the area. It was also affected by patches of contamination from slag or settling lagoons, which in places was severe.

Following closure of the mill, the owners were approached in 1988 by a large retail chain, which entered into an agreement to acquire the whole site. Two planning applications were submitted for a superstore to be located only on that part of the 15 hectares nearest the public highway. No proposal was made for the residual 4 hectares which would have been left with no independent access. The first application was not determined. The second was refused since the local planning authority was anxious to ensure a comprehensive development...

...strategy for the whole site. A third outline planning application in 1988 included B1 and B2 use on the 4 hectare site. Permission was given subject to a complex Section 52 agreement which restricted use of the 4 hectares to B1 and B2 and ensured that once reclamation was completed by the retail chain and access had been provided, the site would be conveyed to an industrial company which operated from the adjacent industrial estate.

7.4 NONINDUSTRIAL PLANNING APPLICATIONS

It was outside the scope of this particular study to monitor the overall loss of industrial land to other uses, through challenges to planning policy. Discussions with the local authorities, however, indicated that such challenges were made, particularly at the height of the property boom. When such cases were contested, applicants would seek to refer to the structural decline in manufacturing industry and the apparent surplus of allocated industrial land. Local planning authorities, on the other hand, would seek to substantiate the longer term perspective of their land allocation and economic development policies.

Most of the challenges to planning policy thus attempted to secure the release of land allocated for industrial purposes for alternative higher value uses. Some of the case studies indicated that opportunities for new industrial investment which occurred at the height of the property boom in 1988—9 were not realized, where landowners were more interested in securing retail permission than in selling to potential industrial users. In most of these cases, neither type of development has yet taken place.

In Crewe, such speculative behaviour was stimulated by the process of local plan preparation, the consultation and deposit stages of which coincided with the property boom. A typical example concerned a ten hectare greenfield site on the edge of Crewe where planning permission for nonfood retail development was refused in 1988. The site, owned by a major manufacturing company, was considered much more suitable by the local planning authority, for the expansion of a nearby industrial estate (Case Study 14). The local authority did concede that one major site in the centre of Crewe, previously allocated for industrial development could in fact be used for retailing (Casc Study 78). It must be recognized in this example, however, that constrained access and the costs of improving the local road network made industrial development unviable. The re-allocation of this site held out the opportunity not only of meeting local retail needs but of resolving the twin problems of an under utilized site and inadequate infrastructure. As discussed in Chapter 3, it will be important to remove sites from employment land allocations where development constraints inhibit their market availability, in order to ensure an appropriate supply.

Case Study 78
Re-allocation of industrial site for retail use

This large site of over 5 hectares in the centre of Crewe was acquired from British Rail by a private haulage company in the mid 1980s and was subsequently used largely for stock car racing. Its development was constrained by poor access. Although there had been plans for many years to improve the surrounding highway network, these had not been implemented . The local authority would not permit development until the necessary road improvements had taken place. The site was previously allocated for industrial use but the costs of improving access could not be met by industrial development alone. This, together with weak market demand, meant that there was no serious interest in developing the site between 1985 and 1988.

The Draft Crewe Local Plan in 1988 reallocated the site for nonfood retail and an outline planning application for retail use, submitted by a private development company, was subsequently recommended for approval, subject to a Section 52 Agreement to ensure the necessary highway improvements were undertaken. The local authority would have derived some planning gain from the proposal, since the developers were to contribute significantly to the cost of the road improvements, enabling this programme of work to be brought forward. In 1990, however, the prospective developer, who had a three year option on the freehold until 1991, went into liquidation and the site therefore remained undeveloped.

Planning permission for residential development was refused on two case study sites, both of which were in the green belt. In one case, planning permission was subsequently sought to use the land for timber storage and car parking in association with the owner's adjacent business (Case Study 79).

In the other case, an established industrial area had long existed as a nonconforming use in the green belt. Planning permission was refused for redevelopment as a new village, but was eventually granted for employment use (Case Study 80). Again, the planning system can be seen here, through a measure of flexibility, to have retained a significant employment site.

Case Study 79
Planning permission refused in the green belt

A 1.6 hectare greenfield site in a Cheshire village was acquired in 1985 by an adjacent company. At that time there were no plans to develop the site and it was kept for future expansion. In 1988, however, two outline planning applications for residential development were submitted by the owner, both of which were refused permission. The site was in the green belt, north-east of Chester. The local planning authority felt that the proposal did not fall within defined exceptional categories of development, and that there were no exceptional circumstances.

Planning permission was sought in 1989 for half of the site to be used for timber storage and car parking in association with the owner's adjacent business. Its turnover had trebled to £3 million per annum and its workforce doubled to 30. To allow further expansion, the company needed to extend its operational area. Permission was again refused and the application was the subject of an appeal in 1990.

Case Study 80
Nonconforming industry in the green belt

The Hooton Employment Area occupied approximately 18 hectares of land in the green belt near Ellesmere Port. Much of the site had been owned by the Ministry of Defence until 1974, when this part was acquired by a well known northern property company. Two smaller parcels were in separate ownerships. The area contained substantial brick buildings dating from World War II and accommodated various employment uses including wooden toy manufacturers, waste recycling, engineering workshops and galvanizing works.

Although this site had been in active industrial use for many years, it had a chequered planning history. In the 1960s and 1970s it was ...

Fig.7.3 Green belt site — planning permission refused (Case Study 79)

Fig. 7.4 Employment site in the green belt (Case Study 80)

...allocated for industrial purposes in the approved Ellesmere Port Town Map as well as being designated as green belt in the submitted West Cheshire Green Belt. In 1977, the draft Cheshire Green Belt Subject Plan identified the great majority of the area as an industrial inset within the green belt. This policy continued to apply until 1985 when the Secretary of State for the Environment, in approving the first alterations to the Cheshire Structure Plan, deleted such insets throughout the county.

In the meantime, many of the buildings within the area had deteriorated and some sites had become vacant or derelict. As a result, the owners appointed consultants to consider the future of the whole area and prepare proposals for comprehensive redevelopment. The property company was at first unwilling, from a financial point of view, to consider a redevelopment purely for employment purposes. A planning application was therefore submitted for the development of a new village, containing approximately 300 houses with associated open space on 14 hectares, with the remaining 4 hectares reserved for 'principally employment orientated uses'. The application was refused on policy grounds in early 1990.

Once the property company acknowledged that the proposed new village stood little likelihood of succeeding in planning terms, the possibility of a redevelopment purely for employment purposes was reconsidered. A planning application was submitted for this purpose in the summer of 1990. At that time, the Ellesmere Port & Neston Local Plan was in preparation and, in the version of the plan placed on deposit in November 1990, the site was 'washed over' by the green belt. The plan, however, allowed existing buildings to be converted to employment uses and provided a measure of flexibility for the redevelopment or extension of buildings. A number of objections were submitted to the detailed policy, as a result of which the local authority amended the deposited plan to provide even greater flexibility in the way in which development and redevelopment proposals could be treated. This enabled the company's planning application to be approved and the site added to the list of available industrial land in Cheshire.

7.5 MAJOR NEW GREENFIELD SITES

No serious attempt was made by the private sector to pursue industrial development, other than in locations acceptable in principle to the local planning authorities. Indeed, none of the agents or speculative developers contacted had contemplated industrial development in such locations between 1985 and 1989. One agent was able to identify a site of 120 hectares in the green belt as a possible future challenge to local planning policy. This, he considered suitable for a distribution depot. Rather than approach the local planning authorities directly, he first intended to sound out INWARD (the agency responsible for attracting overseas investment to the north-west), the Department of Trade and Industry and backbench Conservative MPs.

The research identified one particular site on the edge of Crewe, where the landowner intended to challenge planning policy in the future (Case Study 81). Whenever this proposal comes forward, however, it may well receive the support of the borough council, which, in recent years, has been keen to promote a much higher level of employment development in Crewe.

Case Study 81
Proposed development of unallocated land

This site of 14 hectares lies immediately beyond the built up area of Crewe. Although the Crewe Local Plan, adopted in 1990, proposed that the business park should extend to within a short distance of this particular site, the site itself was not allocated for any development. Nevertheless, the landowner had received a number of approaches from developers and end users, who wished to purchase the site for commercial or industrial development.

Until 1989, the site was let under an agricultural tenancy but this was terminated, with full compensation paid to the tenants. Negotiations with a prospective purchaser were well advanced, when the line of a proposed bypass was announced, crossing the site. This caused the particular purchaser to pull out. Until such time as the exact design of the bypass and, particularly the siting of access points, are known, there is little point in any further discussion with potential purchasers.

In the long term, it will be necessary to seek a change in the planning status of the site, either through a planning application or in the next review of the Crewe Local Plan. Approval will need to be obtained for...

> ...a development sufficiently large to cover the likely high costs of providing internal service roads and overcoming the poor natural drainage and problematic ground conditions present on the site.

Some critics argue that planning restraints only matter in the boom times and at boom locations. Even at Chester, however, where demand is strongest, the city council itself sought to release up to 400 hectares from the green belt, including 57 hectares for new employment uses. As Chapter 3 explained, conservation groups mounted a fierce campaign against these releases and persuaded the Secretary of State for the Environment to intervene. Much to the city and county council's displeasure, the Secretary of State, in approving the Cheshire Replacement Structure Plan in January 1992, reduced the employment land requirement in the Chester District from 210 to 100 hectares. The only real evidence of any serious planning restraint in any part of the Cheshire—Wirral corridor is thus attributable to central rather than local government.

7.6 SUMMARY

There is very little evidence from development control data or from the case studies to suggest that industrial development in the Cheshire—Wirral corridor has been constrained by local planning processes. Indeed all the local authorities were keen to attract additional manufacturing employment and this is reflected in local development plans. Furthermore detailed examination of the sample of 120 case studies shows how planning agreements have been used to bring forward potential industrial land for development and how planning authorities have resisted pressure for the re-allocation of industrial sites for higher value uses, where industrial development was viable. Indeed '*where land or buildings allocated for employment development makes an important contribution to structure plan requirements, there will be a presumption against development for other uses*' (Cheshire County Council, 1992, EMP7).

Where planning restraint has affected the supply of employment land it has tended to be applied by central government, in deleting strategic employment sites in Crewe and Chester from the replacement structure plan and in refusing planning permission for development in the green belt, despite what the local authority appeared to feel were exceptional circumstances. Finally, even where there has been some evidence of local planning restraint, the ability of some development interests to challenge this has been evident and, as in Case Studies 32 and 33, employment considerations have in the end been paramount.

To sum up, flexibility rather than restraint has therefore been the keynote of local planning policy towards industrial development in the Cheshire—Wirral corridor. While such flexibility may well be to the benefit of manufacturing interests, it may act as a deterrent to the involvement of development and investment interests, if they consider that an over supply of development land will result. Some commentators argue that strict restraints on the supply of land at the urban periphery make inner urban development more viable. This view, however, depends ultimately on the willingness of production interests to transfer their investment inwards. In the experience of this study, companies located at the urban periphery or beyond had no such intention, if they were prevented from developing or expanding at their own choice of location, or within two to three miles of that location. From a production viewpoint, flexible approaches in planning and highway matters are thus often necessary to ensure that local industrial investment takes place.

8

PHYSICAL AND INFRASTRUCTURAL CONSTRAINTS

This chapter examines the impact of physical and infrastructural constraints on the market availability of land for industrial development. It summarizes the extent to which 13 specified adverse conditions affected the sample of 120 sites and considers their impact on the development process.

Earlier academic and professional approaches have often concentrated on the existence of physical and infrastructural constraints and have tended to ignore their dynamic nature. In fact, such constraints can melt away in the face of an improving property market or appropriate public sector action. Following analysis of the incidence of adverse site conditions, the chapter therefore explores in some detail the main ways in which such constraints can be overcome.

8.1 THE INCIDENCE AND IMPACT OF SITE CONSTRAINTS

8.1.1 Site size

The size of potential industrial sites may be a constraint upon development, to the extent that size can define the type of development for which a site is suitable. Site size has not been considered a physical constraint in the analysis in the same way as other factors, but it should perhaps be mentioned briefly here. The research did not investigate sites of 0.2 hectares or below. Comparison of the size distribution of the 109 sites either developed between 1985 and 1989 or under construction at the end of 1989 with the 193 included in the potential supply at the end of 1989 suggests that, apart from a possible shortage of sites of less than 1.0 hectares, the size distribution of potential industrial sites was more than sufficient to match previous levels of take up

Table 8.1 Size distribution of industrial sites

	Developed or Under Construction		*Potential Supply*	
	No.	*Proportion*	*No.*	*Proportion*
Less than 1.0 hectares	81	74%	82	43%
1.0—4.99 hectares	22	20%	72	37%
5.0—9.99 hectares	3	3%	19	10%
10.0 hectares and above	3	3%	20	10%
TOTAL	109	100%	193	100%

(Table 8.1). Any future shortage of small sites would be likely to be concentrated in major urban and minor urban areas. In both these locations, the potential supply of larger sites exceeded previous rates of take up. Subject to the willingness of site owners, smaller sites could be created by subdividing larger sites. Thus, size would only be likely to become a constraint if resources were insufficient to fund the necessary infrastructure or if owners were unwilling to subdivide sites.

8.1.2 Physical and infrastructural constraints

The research relied on a subjective evaluation of site conditions. The description of access from the public highway as being 'poor', for example, reflects the perspective of the relevant ownership or development interest. Where possible this evaluation, and judgements about the impact of site conditions on development or marketability, were substantiated by data on the estimated or actual costs of overcoming these conditions, together with information on their impact on the length of the development process if appropriate.

Among the 120 sample sites, poor access, inadequate drainage, adverse ground conditions and the existence of redundant docks, tunnels and other structures, caused the greatest problem in physical and infrastructural terms (Table 8.2). Neither the existence, nor the impact of the constraints were concentrated in any one type of area. Apart from a greater preponderance of docks and structures in inner urban areas and a greater absence of infrastructural connections in rural areas, no noticeable spatial variations existed. Although an adverse site condition was recorded on 186 occasions across 120 sites,

Table 8.2 Physical and infrastructural constraints

Site characteristic	*Number of sites where potential constraint found*			*Number of sites where constraint inhibited development/marketability*		
	Undeveloped (Total 60)	*Developed (Total 60)*	*All (120)*	*Undeveloped (Total 60)*	*Developed (Total 60)*	*All (120)*
Poor highway access	23	6	29	10	1	11
Poor internal site access	21	5	26	2	1	3
Docks, tunnels and other structures	19	7	26	9	-	9
Poor ground conditions	10	13	23	5	-	5
Poor ground drainage	8	7	15	4	2	6
No foul drainage connection	6	6	12	2	-	2
No gas, electricity or telephone connection	4	8	12	1	1	2
Difficult contours	4	7	11	2	-	2
Derelict or obsolete buildings	5	3	8	2	-	2
No surface drainage connection	4	3	7	1	1	2
Contamination of ground	6	1	7	1	1	2
Cellars and substructures	3	3	6	2	-	2
No water supply	2	2	4	1	-	1

only 49 of these were considered to have inhibited the development or marketability of the site. The pattern varies, however, between developed and undeveloped sites.

Only 14 of the 60 undeveloped sites were considered totally free from physical or infrastructural constraints. The other 46 undeveloped sites experienced a total of 115 physical or infrastructural constraints (a mean of 2.5 per site). The development or marketability of the site was inhibited on 42 of these 115 occasions. Of the 60 sites developed or under construction by the end of 1989, 25 had been totally free of physical or infrastructural constraints. The other 35 sites developed or under construction had experienced a total of 71 physical or infrastructural constraints (a mean of 2.0 per site). However, in only seven of these 71 occasions, were the development or marketability of the site affected.

The data suggests that undeveloped sites were more affected by constraints than developed sites. This finding should be treated with some caution. On developed sites, physical constraints had clearly already been overcome and may have been reduced in importance in the mind of the interviewee (or even forgotten). Similarly, on undeveloped sites, what may now appear as a constraint, may in due course be eliminated, and its impact therefore minimized. Indeed, by the end of 1989, action was proposed or had already been undertaken to remedy more than two thirds of identified constraints. Where no action was proposed, it was usually because the owner had no plans to develop or dispose of the site or because the constraint was not felt to inhibit its development or marketability.

Nevertheless, despite the above qualification on its interpretation, the data reinforces the findings of earlier research in Manchester, which suggested that sites remaining vacant were indeed more heavily constrained from those which had been developed (Adams, Baum and MacGregor, 1988). It is also evident that exogenous or off site constraints can be particularly intractable. In a high proportion of the cases where access from the public highway was poor, for example, this was felt to have constrained the site's development or marketability. The resolution of such problems will often require action not only by the relevant development or ownership interest, but also by statutory agencies (Case Study 106).

However, it is equally clear from the data that poor site conditions or inadequate infrastructure do not necessarily inhibit sites being brought forward for development. It is important therefore to consider the circumstances and strategies whereby such constraints will be overcome by the different ownership and development interests identified in Chapter 6.

8.2 CHALLENGING PHYSICAL AND INFRASTRUCTURAL CONSTRAINTS

The distribution between the various ownerships of those sites which experienced at least one physical or infrastructural constraint suggests that production, financial and statutory interests were all prepared to take on constrained sites and progress them towards development (Table 8.3). Indeed, in such cases, the motive for actively owning land in the first place created the incentive to overcome physical and infrastructural constraints. Three key factors were identified which created an environment in which it was possible to address these constraints and which made such action economically viable. These three factors, each of which will be discussed in turn, were increased product demand necessitating expansion, a rising property market producing higher development revenues and the availability of government grants leading to lower net development costs.

Table 8.3 Physical and infrastructural constraints by ownership interest

Ownership interest	*Undeveloped sites*		*Developed stes*	
	No. of sites constrained	*% of sites owned*	*No. of sites constrained*	*% of sites owned*
1. Former landed property	7	100		
2. Production, major manufacturing	10	91	13	57
3. Production, minor manufacturing	2	100	12	63
4. Production, other	4	44	1	50
5. Financial, investment	1	50	-	-
6. Financial, development	2	40	6	86
7. Statutory, local authority	12	75	1	25
8. Statutory, development agency	7	100	2	40
9. Statutory, other	1	100	-	-
TOTALS	46	77	35	58

8.2.1 Increased product demand

For manufacturing interests, the economic upturn of the late 1980s caused product demand to increase or to alter in character. The need for expansionary development was paramount. As a result, investment decisions were often taken at short notice and needed to be implemented as quickly as possible. If land was already owned or could be obtained quickly, the cost and delay involved in tackling difficult site conditions could be programmed into the development process. Companies under pressure to invest or to expand thus regarded physical and infrastructural constraints not as a barrier preventing development but rather as a removable obstacle (Case Study 82). In most cases, there is in fact no urgency on the part of production interests to overcome site constraints until changing production requirements do precipitate development interest (Case Study 83).

Case Study 82
Expansionary development despite difficult contours

Fig. 8.1 Growing industrial company on constrained historic site in a Cheshire conservation area

Although now within a conservation area in a small town in Cheshire, this 0.9 hectare site had been in industrial use since the 19th century. It had been occupied for several generations by a family business, which had expanded, changed and diversified its product range over the years. The site had been used as a mill, a blacksmiths forge, a dairy and now for the manufacture of soft drinks.

The plot was deep, widening to the rear with a relatively narrow frontage on to the main street. Both the factory and a listed building now used as offices, had road frontage. The most recent development was a 740 square metre extension to the rear of the existing production unit, which was completed in 1988. The only notable site characteristic was the difficult contours. In order to keep the levels consistent between the new and original development, it was necessary to move 1500 tons of soil. Grant aid from the Department of Trade and Industry (regional development grant of £61 000 and regional selective...

...assistance of £80 000) was critical to the expansion of this company. Rather than fund the development work directly, the grant aid was also used to offset the cost of overcoming the difficult contours.

Case Study 83
A long term view of site preparation

This 6 hectare site in south Wirral was acquired for expansion in 1979/80 when the purchaser, an international chemical company, anticipated that a new product developed by the parent company in the US would be produced in Wirral for the European market. When it became apparent that the product was not selling well, this plan was abandoned and there has since been no urgency to develop the site.

As a result, reclamation of the site which suffered from poor soil conditions and difficult contours, was not then commenced until 1985 and was expected to take ten years to complete. It was hoped that a new project would be secured for half of the site by 1995, with the other half retained for the company's future expansion needs.

Access to the site from the public highway was very poor and was best achieved through this company's existing plant. Access therefore did not present a constraint on the company's development of the site but could have inhibited its development by any other party. Acquisition by this company in such circumstances therefore represented the best development prospect for the site, but it has remained undeveloped as a result of the company's changing requirements.

8.2.2 A rising property market

A second factor which enabled physical and infrastructural constraints to be overcome was the rise in the property market from mid 1988 to late 1989. This was particularly important to financial interests and to a lesser extent to statutory development agencies and local authorities. One good example of this concerned a land fill site close to the centre of Crewe, which was sold by the borough council to a local development interest in 1989 (Case Study 84). Although in a good location, ground conditions made it unmarketable until the late 1980s, when the property market made its development economically viable.

Case Study 84
Market recovery helps overcome physical constraints

Fig. 8.2 Successful speculative development of severely constrained site (Case Study 84)

Site 84 was a level, filled plot of 1.3 hectares with good access to Crewe town centre. The ground conditions were so poor, however, that one local property agent remarked that he *'would not have touched it with a barge pole'.* The site had been owned by the local authority for many years but was sold to a private development company in 1989.

Prior to the construction of 12 small industrial units, the developer had to strip out the existing fill and replace it with new compacted material. A large water mains which crossed the site had to be relocated at a cost of £70 000. To an outsider, site preparation would have appeared to have been a prohibitively expensive operation. However, average industrial land prices in Crewe which had remained...

...static at approximately £85 000 per hectare between 1985 and 1987, rose to £100 000 per hectare in 1988 and leapt to £225 000 per hectare in 1989. For this site, the developer paid £110 000 per hectare in 1989. This provided the developer with a substantial discount towards the cost of tackling difficult site conditions. Phase 1 of the completed development was marketed at a competitive rent in 1990 and the six units were quickly sold. Development of Phases 2 and 3 then commenced.

8.2.3 Public sector financial assistance

Unlike private production and development interests, the activity of local authorities and statutory development agencies in preparing sites and constructing units is less dependent on the economic cycle and property market trends. That is not to say that market appraisal is not an important component of their economic development strategy. Quite the contrary would be true, but as already illustrated in Chapter 6, these agencies play a key role in site assembly, reclamation and even development, in anticipation of market demand and to fill market niches which are less attractive to private developers. This activity is partly financed not only by their own development budgets but also by grants from central government and European funds.

All the ten grants for those sites which remained undeveloped in 1989 for example, had been obtained by statutory agencies. Indeed, in the Cheshire—Wirral corridor, only local authorities, English Estates and the Merseyside Development Corporation were prepared to tackle problematic physical and infrastructural constraints, in advance of the prospect of development. Their role in overcoming supply constraints and in making land genuinely available was critical. The discretionary nature of grant funding, however, means that substantial staff time must often be devoted to formulating applications. One authority which has been particularly successful in obtaining both European Regional Development Funding and derelict land grant has been Ellesmere Port & Neston Borough Council. Complex funding packages have been assembled to promote the successful development of often very difficult sites (Case Study 85).

Case Study 85
Grant package enables reclamation and development

Fig. 8.3 Developed site in Ellesmere Port which had once experienced multiple constraints

This 1.4 hectare site was brought forward for development by the local authority entirely as a result of the use of financial assistance. The site, which was previously in two separate ownerships, experienced considerable changes in level partly as a result of tipping. There was also toxic contamination and a lagoon. It was acquired by Ellesmere Port & Neston Borough Council in 1980/81, at a total cost of £25 000, and was cleared and levelled over the next two financial years at a further cost of £450 000. Acquisition and reclamation were funded by derelict land grants, as was the subsequent cost of £78 000 for piled foundations in the later phases of development. Construction began in 1985/86 and the final phase, Phase 4 comprising five units each of 100 square metres, was completed in June 1990. For the first two phases, an urban programme grant of £200 000 covered 75% of the development costs. ERDF grants were applied for retrospectively for the final two phases. A total of 19 small industrial units were provided and all were successfully let.

Some local authorities were concerned that the rules inherent in derelict land funding only allowed them to undertake basic reclamation and infrastructure provision. The high standard of landscaping necessary to attract quality industrial development had to be funded from other sources. One particular irritation for local authorities was the claw back regulations on successful derelict land schemes. Where reclamation increases site value (as it normally does with hard end uses) then 50% of the enhanced value must be repaid up to the total amount of the derelict land grant. This, in the view of Ellesmere Port & Neston Borough Council, for example, failed to reward successful local authorities, particularly as the funds clawed back were returned directly to the Treasury rather than being made available again to the local authority concerned or even added to the following year's funding for the region as a whole.

The importance of financial assistance in overcoming physical constraints is not limited of course to statutory authorities. Although production growth and the need to expand may provide production interests with the incentive to overcome constraints, and the increased revenue may underwrite part of the costs of expansionary development, it may be economically viable in some cases only if financial assistance is available. Indeed a number of examples have been given of developments where grant aid was critical (Case Study 86).

Case Study 86
Regional assistance critical in stabilizing site conditions

This site of 15 hectares in Wirral had been owned and occupied freehold by the same chemical company since 1850. The whole of the site, apart from 1 hectare, is now in operational use. If necessary, additional land adjacent to the boundary could be acquired from another company in the same group.

In 1985—6 the company obtained regional assistance from the Department of Trade and Industry towards the cost of two new process buildings, with a total floorspace of 3000 square metres. The buildings were constructed on cleared land, once occupied by redundant sheds which had previously been demolished into their cellars.

Prior to construction of the new buildings, the cellars had to be dug out and, since the site was made ground, piled foundations needed to be installed. Although it took only four weeks to stabilize ground...

...conditions, the work added £30 000 to development costs. Regional assistance was given nominally to cover the standard proportion of the allowable costs of construction plant and machinery. This assistance, however, was instrumental in persuading the company to meet the high costs of preconstruction works, caused by the problematic physical conditions. Since regional assistance is now given on a discretionary basis, and related closely to the cost per job, it is unlikely that these developments would have taken place on the same site, under the revised rules.

Similarly, private speculative development even in a relatively buoyant property market may be economically viable only where development costs are off set by financial assistance. In the Cheshire—Wirral corridor this was particularly true in the inner areas of Wirral where rental levels were lower than would be necessary in some cases to justify new development or even the refurbishment and conversion of obsolete buildings (Case Study 54). One dilemma faced by developers, however, is programming the project to allow for the bureaucratic processes involved in securing financial assistance. One developer estimated that the interest charges incurred while his grant application was being processed exceeded the eventual amount of aid received (Case Study 116).

In theory it could be argued that where physical and infrastructural constraints exist, the costs of overcoming them should be reflected in lower land values and that grant aid should not be necessary in these circumstances. In reality, earlier research (Adams, Baum and MacGregor, 1988) has shown that although there is indeed some adjustment, land prices do not fully reflect the cost of remedial action. This argument could not apply however to those cases where in situ expansionary development depended on financial assistance, expansion which may secure existing jobs and afford employment growth. Such grant aid will therefore continue to be a necessary instrument in achieving regional and local economic objectives.

The research revealed that 24 of the sample sites now developed or under construction and nine of those undeveloped, had received one type of grant or another. Some sites received more than one grant. In fact, a total of 41 grants were provided towards the reclamation, servicing or development of the sample sites (Table 8.4). It should be borne in mind in interpreting these figures, that the sample included an above average proportion of sites where reclamation had been undertaken or assisted by the public sector. However,

Table 8.4 Reclamation development and infrastructure grants by ownership type

Ownership interest	*Undeveloped sites*	*Developed sites*
1. Former landed property	-	-
2. Production, major manufacturing	-	8
3. Production, minor manufacturing	-	11
4. Production, other	-	1
5. Financial, investment	-	-
6. Financial, development	-	2
7. Statutory, local authority	9	7
8. Statutory, development agency	1	2
9. Statutory, other	-	-
TOTALS	10	31

a recent survey in Greater Manchester also discovered that 46% of industrial land where development was completed or due to be completed between 1986 and 1991 depended on public sector grants to bring it into a developable state (Roger Tym & Partners, 1989).

Reclamation, development and infrastructure grants came from a variety of sources, including the local authorities, the Department of the Environment, the Department of Trade and Industry, and the European Community. Regional development assistance was most numerous (14 cases), closely followed by derelict land grant (12 cases). Three cases were recorded where urban development grant (or its equivalent) was awarded, but none for its successor, city grant. Agencies which received grant assistance considered it crucial to the prospect of site development in almost 75% of cases where it was received. Derelict land grant was considered critical in 92% of cases where it was awarded but regional development assistance was regarded as critical in only 64% of cases. The abolition of the regional development grant in 1988 now means that all regional assistance is on an entirely discretionary basis.

Particular grants could not necessarily be assigned to either reclamation, development or infrastructure. European Regional Development Fund, for example, covered a proportion of both infrastructure and development costs. Derelict land grant often covered not simply the cost of reclamation but also the provision of basic services and landscaping. In some cases, derelict land grant was only needed to fund the servicing costs of an overall package (Case Studies 7 and 8). More importantly, within the private sector, grants occasionally proved critical for purposes other than that for which assistance was given (Case Studies 82 and 86). In the private sector, such virement of funds could be disguised in internal accounting procedures. A later section in Chapter 10, however, will discuss how the rules governing some grant regimes may in fact delay or inhibit development.

Manufacturing companies were active in their pursuit of grant assistance, but like the private sector concentrated their activity on sites about to be developed. This may be partly attributable to the rules and regulations surrounding grant assistance. Derelict land grant, for example, can only be obtained by the private sector at 80% of eligible costs in comparison with 100% for local authorities. When private sector derelict land grant was incorporated within city grant in the 57 partnership or programme areas in 1988, it further deterred reclamation by the private sector, in advance of a specific development proposal.

8.3 UNDEVELOPABLE SITES

Most physical and infrastructural constraints identified in the research already had been overcome, or were likely to be tackled in the future. In almost all cases, therefore, it is much more healthy to concentrate on the speed and likelihood of overcoming such constraints rather than on their existence. However, a hard core of four out of the 120 sample sites were, to all intents and purposes, undevelopable through a combination of physical and ownership constraints. These four sites should have been long deleted from any register of available industrial land kept by the local authorities (Case Studies 87—90). In Wirral, in particular, the development of such sites cannot be envisaged while there is a sufficient supply of vacant urban sites with less severe multiple constraints.

Case Studies 87, 88, 89 and 90
Allocation of undevelopable sites

Sites 87—90 were identified for industrial development by Cheshire County Council or Wirral Metropolitan Borough Council but were so constrained by ownership and physical constraints that they had...

...no real development potential, at least for the foreseeable future.

Site 87 in a rural area near Chester, comprised 0.2 hectares. Although used as sidings until the 1970s, it had been cleared for at least five years. It was an extremely narrow site, and could only be accessed across adjacent land in a separate ownership. Significantly, it was being held by British Rail for future operational use, in case the nearby line was electrified. In the meantime, the site was let on a one year licence for the repair of JCBs and lorries.

Site 88 in the inner urban area of Birkenhead was 0.9 hectares in area and was the residual land remaining following the earlier development of the bulk of a site of over 4 hectares. The whole site was acquired in 1981 from British Rail by English Estates who developed workshops and factory units in three phases, the last of which was completed in 1985. Again the residual area comprised a narrow plot, which was additionally constrained by its proximity to a residential area and by particularly adverse ground conditions. This area was unable to take a load without the developer incurring very substantial costs. There had therefore been no attempts to market the site which was tenanted on a short term lease, and was used as a scrapyard.

Case Study 89 was 0.35 hectare site in Wirral docks, identified by Wirral MBC in its survey of potential employment land. However, it appeared to have no possibility of being developed for industrial use. It had very poor access to the public highway, being close to a set of traffic lights at a four way junction and close to a bridge. In addition it sloped down to the dock and would be difficult to level. As a result of these constraints, Mersey Docks & Harbour Company never had any plans to develop the site, although the adjacent berth was operational. Neither had they marketed it for industrial use, in view of its limited potential. In 1990, the site was leased to a rowing club.

The identification of Site 90 by Cheshire County Council in its land availability records was highly optimistic, even though it was acknowledged not to be 'immediately available'. It was a difficult triangular area of 0.54 hectares, sandwiched between a canal and the M53 in Ellesmere Port. Owned by British Gas, it was once part of a larger site on which there had been a gas reforming plant which was operational until 1987. The rest of the site was sold to the local authority. Development of Site 90 was constrained not only by its shape, but more particularly by three high pressure gas pipes on site,...

...the easement for which reduced the developable area by half. The type of uses possible on the site were also limited by these pipes. In 1989, it was let on a short lease to a manufacturer of timber garden buildings. The lease stipulated that any storage on site should be limited to those items which two people could move in 15 minutes. The prohibition of activities which involved noxious substances or attracted numbers of people was further indication of the site's physical constraints.

British Gas had not actively marketed Site 90, neither did it intend to undertake development itself. Industrial values did not make development here viable given the shape of the site and the cost of relocating the pipes. The owner was therefore content to derive a rental income from the site.

8.4 SUMMARY

The evidence from the case studies is that many sites experienced physical and infrastructural constraints, and that many sites experienced multiple constraints. Furthermore, it would appear that undeveloped sites experienced more physical constraints than developed sites. Such characteristics, however, do not always inhibit development or prevent land exchange and those structural conditions which enable such sites to be brought forward by the various development interests have been illustrated. The three key factors have been economic growth, property market trends and the availability of financial assistance to the principal players—statutory authorities, production interests and private development interests. Furthermore, the analysis has shown that it was not necessarily easier to bring land forward for industrial development in rural and minor urban areas than in major or inner urban areas. The process of development therefore requires appropriate treatment on a site by site basis, irrespective of area.

9

MARKET ACTIVITY AND DEVELOPMENT PROSPECTS

It is evident from the analysis and many case studies in earlier chapters that the industrial land market in the Cheshire—Wirral corridor does not operate according to neo-classical theory. Land ownership and development interests are diverse. Land itself is not homogeneous. The market in fact comprises a number of submarkets, whether defined spatially, by size of site and tenure, or by the other specific needs of particular development interests. Furthermore it has been shown how a number of different development constraints may inhibit the immediate availability of industrial sites. These include national planning objectives, physical site characteristics and ownership behaviour.

The work of Fothergill and colleagues discussed in Chapter 1 and elsewhere drew attention to a relationship between supply constraints on land in existing urban areas and industrial decentralization to small towns and rural areas. One of the paradoxes of the inner urban land market, however, which defies neo-classical theory is the apparent coexistence of a shortage of sites and extensive land vacancy (Adams, Baum and MacGregor, 1988). Aggregate data alone may shed little light on this phenomenon. The locational characteristics of industrial land are of prime importance, particularly in areas where there is little inward investment and low levels of speculative development. In these circumstances, available land may remain undeveloped unless it is adjacent to existing factories wishing to expand. A number of examples have been given where existing users have then gone to tremendous lengths to develop even heavily constrained sites rather than relocate (Case Study 31). Conversely sites which are entirely free from constraints are not necessarily developed.

Chapter 6 illustrated the different land management strategies of the principal ownership interests identified in the Cheshire—Wirral corridor. It has been shown how local planning policy has responded positively to the needs of industrial development (Chapters 3 and 7) and how physical site constraints may be overcome by cyclical and structural factors (Chapter 8).

It is clear, therefore, that ownership behaviour may be a critical determinant of the effective supply of industrial land. Within a flexible planning environment, development interests are willing to overcome site constraints if ownership interests are responsive to market demand and if land is made available on terms and conditions which are favourable to development.

This chapter will therefore examine in more detail the relationship between land ownership, land exchange and land development for industrial use. It will determine which ownership interests contributed actively to the process of making land available to industry. It will also describe how ownership behaviour constrains the supply of land and will specifically illustrate how passive ownership may be challenged, drawing together arguments from earlier chapters and examining the role of statutory agencies and of compulsory purchase procedures.

9.1 ACTIVITY IN THE LAND MARKET

A total of 50 transactions were recorded for the 120 sample sites. Undeveloped sites accounted for 18 of these transactions, and sites developed or under construction for 32. Two types of vendor were dominant: local authorities and major manufacturing interests (Table 9.1; Fig. 9.1). Together, they accounted for almost 60% of all sales. Almost half the purchases were made by minor manufacturing interests and private development interests. Major manufacturing interests and statutory development agencies were important purchasers, but to a lesser extent. The flow of sites mirrors what has already been established about the key role of the statutory sector in site assembly, reclamation and servicing for subsequent industrial development and in particular the different patterns of landholding by major and minor manufacturing companies to meet future needs.

Transactions were often an essential prerequisite to development, particularly for minor manufacturing interests, financial development interests and statutory development agencies. The proportion of sites developed by each of these interests which were owned by the same interests four years prior to the commencement of development were respectively five out of 19, one out of seven and zero out of five. In contrast, 17 out of the 23 sites developed by major manufacturing interests were owned by the same interest four years prior to the commencement of development, and in most cases had previously been retained for subsequent in situ expansion.

Table 9.1 Transactions in sample sites between ownership interests

Vendor	*Purchaser (categories defined as for 'vendor')*									
	1	*2*	*3*	*4*	*5*	*6*	*7*	*8*	*9*	*Total*
1. Former landed property	-	-	-	-	-	1	-	-	-	1
2. Production, major manufacturing	-	5	1	5	-	-	1	2	-	14
3. Production, minor manufacturing	-	-	1	1	-	2	-	1	-	5
4. Production, other	-	1	2	-	-	3	-	1	-	7
5. Financial, investment	-	-	2	-	-	-	-	1	-	3
6. Financial, development	-	-	1	-	-	1	-	-	-	2
7. Statutory, local authority	-	-	7	-	-	2	3	3	-	15
8. Statutory, development agency	-	1	-	-	-	-	-	-	-	1
9. Statutory, other	-	1	-	-	-	1	-	-	-	2
TOTALS	-	8	14	6	-	10	4	8	-	50

Note: In contrast to Table 9.1, Table 6.1 identified only the purchasers in the 50 transactions.

Companies without their own reserves of land were dependent on transactions in order to expand. Few companies appeared to undertake the advance purchase of additional land in the vicinity, as and when it became available, although where relocation of a company would present particular difficulties this can clearly be advantageous (Case Study 91).

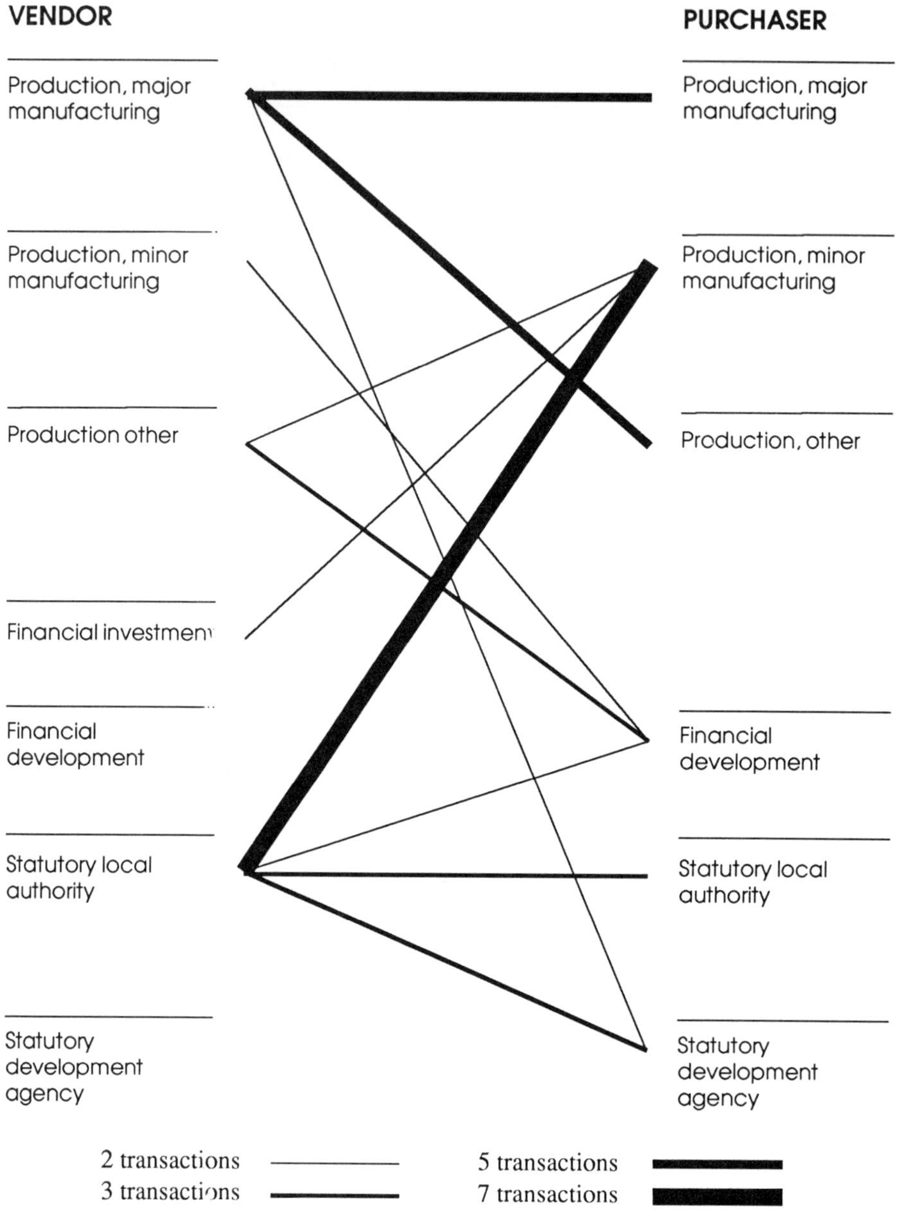

Note: This figure illustrates the main flows described in Table 9.1.

Fig. 9.1 Main transaction flows for sample sites

Case Study 91
Advance purchase of expansion land

The mobility of a local engineering company in Wirral was heavily constrained by its dependence on a dry dock, leased from Mersey Docks & Harbour Company and located about a mile from the main site. The opportune purchase of land adjacent to its main site was critical in allowing the company to extend its production facility, when needed.

In 1986, a concrete batching plant occupying a site of 0.24 hectares in inner Birkenhead closed, as a result of a company merger. Within a month, and prior to any formal marketing, the site had been acquired for £18 250 by Gould and Parry, the adjacent company. Gould and Parry are a long established marine and chemical engineering company who employed over 300 people in Wirral and subcontracted work to many other specialist firms. When the site was acquired in 1986, Gould and Parry had no plans for its development. The closure of the concrete batching plant was, in fact, too good an opportunity to miss, and the site was acquired as quickly as possible both as an investment and to open up the possibility of future expansion.

Two years later, Gould and Parry took advantage of their opportune purchase, when growth in the company's business justified the extension of their premises. Plans were drawn up for the construction of new offices and the extension of the existing machine shop onto the new site. A total of 1500 square metres of new floorspace was proposed. Apart from the need to demolish some of the remaining structures associated with the concrete plant, there were no physical constraints to development. However, development was delayed while Gould and Parry sought financial assistance from three public sector sources. A low interest loan of £371 000 was received from the European Community and a sum of £90 000 in regional assistance was obtained from the Department of Trade and Industry. When development finally commenced in December 1989, a decision was still awaited on a grant application to the Merseyside Development Corporation to cover the cost of some of the external works. Although financial assistance was considered extremely helpful, the company admitted that, since the growth of the business necessitated extensions, development would have gone ahead even without such assistance.

Table 9.2 Owners intentions for undeveloped sites at end of 1989

	Disposal		*Retention*			*Total*
Ownership interest	*Under offer*	*On market*	*To be marketed*	*To develop*	*No plans*	
1. Former landed property	2	3	1	1	-	7
2. Production, major manufacturing	-	1	5	1	4	11
3. Production, minor manufacturing	-	-	-	2	-	2
4. Production, other	-	1	1	2	5	9
5. Financial, investment	2	-	-	-	-	2
6. Financial, development	-	1	-	4	-	5
7. Statutory, local authority	5	5	5	1	-	16
8. Statutory, development agency	2	1	2	1	1	7
9. Statutory, other	-	-	-	-	1	1
TOTALS	11	12	14	12	11	60

The analysis in Table 9.1 focussed on actual transactions for the sample sites. Separate analysis of the 60 sites which remained undeveloped at the end of 1989 revealed that only 23 were then under offer, on the market or known to be available (Table 9.2). The remaining 37 were, for one reason or another, held off the market by their owners. Although 37 of the 60 undeveloped sites were held off the market at the end of 1989, it would be a mistake to equate this with an ownership constraint or even to consider that the landowner alone was responsible, in every case, for any delay to development. In fact, 26 out of the 37 sites, retained by their owners at the end of 1989 were intended to be marketed or developed in due course. In some of these cases, there was good reason not to market sites immediately. In particular, landowners may not consider it worthwhile to actively market sites where constraints severely inhibit their development potential. They may nevertheless be willing to release them if approached by prospective purchasers (Case Study 92) or may intend to develop the sites eventually themselves (Case Study 93). In many other cases, work was in progress to prepare the sites for marketing or development by undertaking reclamation, improving access or establishing title. In early 1990, development commenced on three of the 26 sites and marketing on a further two.

Clearly historic patterns of land ownership and current market behaviour do not necessarily indicate the existence of an effective ownership constraint. The following section therefore offers a definition of active and passive behaviour, before going on to explore how ownership behaviour may constrain the supply of land in various ways.

Case Study 92
No active marketing of landlocked site

Site 92 had been owned by a major manufacturing company for many years. It was undeveloped but had not been actively marketed. Because of a highways constraint, it had not been considered worthwhile to market it in isolation, and the case study is one example where perceived ownership behaviour cannot be interpreted as an ownership constraint.

The only access to this 0.8 hectare site in Bromborough, Wirral was afforded by a narrow track and, as a result, the site was effectively landlocked. It had development potential only in association with either of the two adjacent sites. One of these was operational and Site 92 might therefore have provided expansion land at some point. The other was the 12 hectare site of a former chemical works, acquired by developers in 1988, cleared and subsequently the subject of a planning application in 1990 for B1, B2 and B8 use. This developer had expressed interest in acquiring Site 92 since it would provide the proposed business park with a river vista.

Case Study 93
Constraints on marketing

Fig. 9.2 Local authority depot, for sale in 1993

This site of 2.7 hectares in Bromborough was a former municipal depot, which the local authority had declared surplus in 1988. It was then leased on a temporary basis for a waste transfer station. In 1989, the local authority obtained outline planning permission for B1, B2 and B8 uses and had intended to market the site freehold in spring 1990.

In fact, it proved much more difficult than anticipated to relocate the tenant and marketing could not commence as planned. Although the tenant would have been happy to accept a much smaller office building together with only 0.4 hectares for storage and vehicles, very few such suitable sites existed for the relocation of 'messy uses'.

From an employment view, the local authority was unwilling to evict the tenant until a suitable relocation site was found. In the meantime, marketing was shelved and, although nominally available for development, the site remained under used.

9.2 ACTIVE AND PASSIVE BEHAVIOUR

It is important to conceptualize how the various ownership interests relate both to the process of development and to the operation of the land market in the Cheshire—Wirral corridor. This conceptualization is illustrated in Fig. 9.3 and is based not only on the analysis of actual transactions (Table 9.1) and of the future intentions of the owners of undeveloped sites at December 1989 (Table 9.2), but also on an evaluation of the land management behaviour of the different interests prior to these two points in time. Land consumers are subdivided into those industrial users who required land for production purposes and those development and statutory interests whose prime motive in acquiring land was to undertake reclamation, servicing or speculative development. Land suppliers are subdivided into those whose behaviour was demonstrably active and those whose behaviour was more passive (Adams and May, 1991).

Active suppliers are those who placed their land on the market or made its availability known. Active suppliers often took necessary action to move their land forward for development. In many cases, this merely involved applying for planning permission in advance of marketing. In other cases, however, action was taken to overcome problematic physical or infrastructural constraints in order to render the site more marketable. **Passive suppliers**, on the other hand, were those who took no particular steps to market their sites, even though they may have had a distant intention to do so. Often they merely responded, or failed to respond, to offers from land consumers. They tended to make very little effort to prepare sites in advance for development, leaving any necessary action to prospective purchasers.

In the Cheshire—Wirral corridor, certain interests related to the industrial development process and to the industrial land market in particular ways. Five of the interests tended to adopt only a single role. Former landed property and 'other statutory interests' (an area health authority and the British Rail Property Board) acted primarily as passive land suppliers. In four out of the five cases where undeveloped sites owned by former landed property were either under offer or on the market at the end of 1989, the decision to sell had been initiated by an approach from a development interest. In the fifth case, the site was only released as a result of the death of a landowner and the response by his heir to the changing economics of a large agricultural estate and the need for diversification (Case Study 94).

Minor manufacturing interests were primarily land consumers for their own use. In the 50 identified transactions, minor production interests had been the purchaser on 14 occasions. At the end of 1989 they retained only

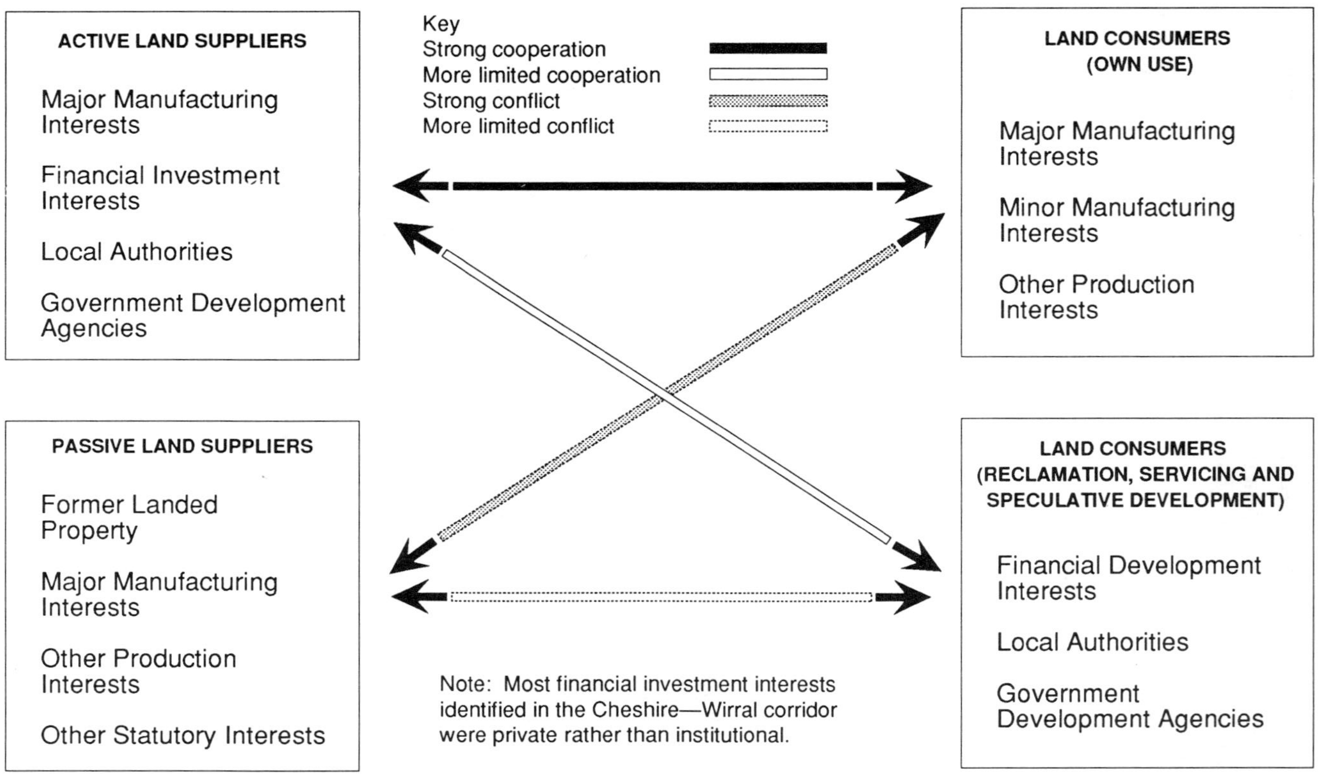

Fig. 9.3 Land suppliers and consumers

two undeveloped sites which they did intend to develop. When minor manufacturing companies were the vendors of sites, these sites tended to be released only in response to direct approaches from other companies or as a result of bankruptcy or the need to relocate. Financial development interests were interested in site acquisition principally in order to build for industrial users. The local and regional development companies active in the Cheshire—Wirral corridor appeared not to hold land speculatively and did not emerge as significant suppliers of development sites. Private development interests intended to develop four of the five undeveloped sites which they held at December 1989. They had been the purchaser in ten of the identified transactions and the vendor in only two cases. Finally, as might be expected, given their motive for land ownership, financial investment interests appeared as active land suppliers.

Case Study 94
Inheritance activates development

This case study is an unusual example of active site ownership by a landed property interest. Traditional land management strategies were transformed by economic factors. This was only made possible, however, by a transfer of ownership.

The site was part of a large agricultural estate owned by the same family since around 1850. Over the years the number of individual farms on the estate had fallen from 100 to around 20. This rationalization led to a number of farm buildings becoming redundant. In the case of this 1.1 hectare site, there was perhaps 2000 square metres of built space which was now derelict and would have to be demolished in the course of any redevelopment.

The site was in fact allocated for employment use in 1985 in the Chester Rural Areas Local Plan, but had remained undeveloped. The reason for this lay in the priorities of the owner who died in 1989. His main concern was to see that the traditional character of the estate remained unchanged and that available land was fully utilized for farming. In 1990, however, his son hoped to diversify by putting under utilized or derelict sites to commercial use. The possibility of granting a 99 year lease to a construction company in order to redevelop Site 94 was under discussion. This strategy would generate long term income which would be reinvested in agricultural plant and buildings.

Four interests clearly adopted more than one role. Major manufacturing interests were a key player in the industrial land market, being involved in 17 of the 50 identified transactions. They acquired eight sites and released 14 on five occasions to other major manufacturing interests. Major manufacturing interests behaved in three different ways. In many cases, corporate restructuring and industrial decline meant that major manufacturing interests had substantial areas of surplus land. Such manufacturing interests were thus able to act as land suppliers. Some took to this role enthusiastically and were keen to release land where exchange values were high, but the majority were much more passive in their behaviour, only slowly coming to the realization that their surplus sites would be unlikely to be required for their own purposes again. Of course, in many cases, major manufacturing interests were also purchasers, acquiring sites both for new and expansionary development.

The behaviour of 'other production interests' was primarily passive. They intended to retain the majority of undeveloped sites identified in their ownership at the end of 1989, often with no plans for their disposal or development. This element of the sample was dominated, however, by the sites owned by the Mersey Docks & Harbour Company and Manchester Ship Canal Company and their land management and development strategies, discussed in more detail in Chapter 6. However, in certain cases, land was required to help maintain or regenerate these activities. Local authorities and government development agencies also played important roles, the first particularly as active land suppliers and the second as purchasers in order to both reclaim, service and release sites and also to construct advance units to let.

9.3 OWNERSHIP BEHAVIOUR AND SUPPLY CONSTRAINTS

The discussion above has defined active and passive ownership behaviour and has suggested the principal roles played in the land market and development process by the various ownership interests identified in the Cheshire—Wirral corridor. The behaviour of different interests, however, can only be said to give rise to a constraint either if existing market demand does not induce supply or if sites are only made available on terms and conditions which are not favourable to development. This section will consider the extent to which landowners' expectations and land management strategies gave rise to supply constraints.

9.3.1 Valuation constraints

Valuation constraints occur when transactions in industrial land are held up because buyers and sellers cannot agree on price. Such disputes are usually caused by uncertainty in the industrial land market, particularly in relation to the extent of price movement. Previous research has suggested that valuation practices can delay development, especially in inner urban areas. In fact, only three of the 120 sample industrial sites in this research were affected by valuation disputes, of which only one was located in the inner city (Case Study 95).

Case Study 95
Market uncertainty in the inner city

In this case study, completion of the purchase of the site was significantly delayed by protracted and difficult negotiations over the price, at the end of a period when the market had been relatively dormant.

This site of 0.2 hectares in inner Birkenhead was acquired by a plastics reprocessing company in 1987. When the company's existing premises in Wirral docklands were threatened by the plans of Merseyside Development Corporation, this alternative site offered an opportunity to relocate to a larger and better quality unit. Approximately 40% of the site was covered by an existing building of about 700 square metres. The premises and location were considered ideal by the reprocessing company and no alterations other than the installation of an electricity substation and the digging of pits for machinery were required. The company was therefore prepared to tolerate relatively poor access on to the public highway, with lorries able to turn only in one direction on to the main road.

Negotiations to acquire the site took ten months to complete, as a result of price uncertainty in this part of the inner area. An initial offer of £55 000 was accepted by the vendors, but shortly afterwards an asking price of £80 000 was set instead. Months of negotiations produced an eventual settlement of £65 000 for the site and buildings, which was in fact £5000 less than the valuation undertaken by the purchaser's bank. The purchaser spent an additional £75 000 installing the substation and machinery for which the company received a...

...contribution of £1200 in regional development grant.

The suitability of this site in close proximity to existing premises made the purchaser determined to see negotiations through to a successful conclusion, despite the protracted uncertainty over the purchase price.

In the two other cases no agreement was reached and the deal was called off. One of these concerned the sale of land by former landed property to a major manufacturing company. The site, of just over four hectares, was in agricultural use and, up until 1985, was included in the North Cheshire Green Belt. It was excluded from the green belt specifically to allow the adjacent manufacturing company to expand. Planning permission had actually been given for this purpose (Case Study 96). The third example concerned a greenfield site of 2.3 hectares owned by Wirral Metropolitan Borough Council (Case Study 97) which experienced severe physical constraints. The value placed on the site by the prospective purchaser was considerably less than the local authority's more optimistic valuation, which in part reflected its own financial constraints and objectives.

Case Study 96
Valuation constraint in rural area

This level site of just over 4 hectares, which was in agricultural use, was designated as an employment inset in the green belt in the Chester Rural Areas Local Plan in 1985. This was specifically to meet the expansion needs of the major manufacturing company established to the east of the site. Indeed planning permission had been given to the company in 1984 for an industrial extension on the site. This, however, has not been implemented because negotiations over the purchase of the site broke down in 1988.

The site had no road frontage and was accessed from the highway by a farm track. Access could be achieved by the major manufacturing company from within its existing site. Since the manufacturing company considered itself the only likely buyer, it offered only £75 000 per hectare in comparison with mean values of £125 000 per hectare, prevailing in the locality at the time. On top of this, the manufacturing...

...company sought a further reduction of £50 000 on the total price to reflect the cost of relocating a foul sewer which crossed the site.

The offer proved unacceptable to the vendor, a family trust which felt no urgency to dispose of the site. After 12 months, negotiations thus broke down and the land remained undeveloped. The trust's agents intended instead to make representation during consultations on the proposed structure plan alterations. If the allocated area could be extended to have its own road frontage, more general marketing would become worthwhile.

Case Study 97
Unrealistic view of market value by landowner

There can be a tendency on the part of owners of severely constrained land to concentrate on its enhanced value after reclamation or preparation and to ignore its likely negative value prior to such work. This greenfield site of 2.3 hectares, owned by Wirral Metropolitan Borough Council, represented almost the last available plot on an industrial estate in north Wirral. Owing to its proximity to a river, ground conditions were extremely poor and piled foundations would be required. Almost £1 million would need to be spent to bring the site up to a developable standard.

In 1989, English Estates expressed interest in constructing small industrial units on 1.8 hectares of the site. An asking price of £125 000 per hectare was suggested by the local authority. Up to £100 000 might have been returned to the purchaser by the local authority in the form of a grant towards alleviating problematic site conditions. In the view of English Estates, however, the site was worthless.

At the time of the negotiations, the local authority was in the midst of a severe financial crisis. The authority's financial objectives had priority over its economic development policies. The site remained undeveloped and, in the circumstances, was probably suitable only for open storage or for a reversion to agricultural use.

In the vast majority of cases, however, transactions in industrial land took place without any serious dispute arising over the price of the land. The small number of valuation constraints can be attributed to the role of manufacturing companies rather than speculative developers and investors as the main development agents in the corridor. Manufacturing interests were principally land price takers. They tended to assess the future use value of the site rather than pay too much attention to the exchange value at which it was offered. Previous work (RICS, 1978) suggested that land costs, at that time, accounted for approximately 11.5% of total costs in the case of industrial and commercial development. The case studies indicated that, since manufacturing companies assessed project viability in business rather than in development terms, they were unlikely to quibble over the price asked for land, unless it seriously jeopardized business viability.

9.3.2 Marketing strategies and terms and conditions of disposal

Throughout this analysis, but particularly in Chapter 6, examples have been given of the ways in which the supply of industrial land might be influenced by the wider objectives of ownership interests. In particular it has been shown how the terms and conditions on which land is made available may be unattractive to development interests and can inhibit the take up of potential industrial sites. This applied particularly to local authorities concerned to achieve their economic development objectives and anxious to prevent speculation in land. In some cases, this has considerably delayed development (Case Study 98). In others, onerous leasehold agreements delayed the transaction to the point where a downturn in the economy made development unlikely for the foreseeable future (Case Studies 63 and 64).

Case Study 98
Restrictive marketing by local authority

In principle, this 1.1 hectare site, owned by Crewe & Nantwich Borough Council, had been available for over ten years. Having a prominent location at the entrance to a modern employment park, however, the local authority felt it to be suitable for hotel development, for which there was an identified need. No firm proposals came forward and when discussions began with a hotel chain regarding the construction of a hotel complex on the new business park, the site identified here was released for industrial development. Owing to location, however, the local authority remained concerned to ensure a prestigious and attractive development and was not prepared to...

...sell the land to a speculative developer, despite considerable interest from development companies.

In 1989, planning permission was given for a 3000 square metre development which comprised a bakery, offices and ancillary buildings. It did not progress, however, as the costs of relocating plant and of developing the site, estimated to be £1.5 million in total, could not be justified. This was disappointing to the local authority since it was an attractive proposal made by a local company. In autumn 1990, the site was under consideration by a well known American toiletries company seeking a European base. The proposed development would again comprise a mix of manufacturing and office space and continued to reflect the local authority's criteria with regard to this site.

The evidence suggests that these problems, however, are not intractable. Of greater concern is the marketing of potential industrial sites for higher value uses. Examples were given in Chapter 4 of pressure to release land for retail and housing development, especially during local plan preparation. This strategy particularly applied to land held by major manufacturing companies and financial interests. By the end of the 1980s, however, the marketing strategies even of local authorities were increasingly influenced by financial constraints rather than employment priorities (Case Study 99).

Case Study 99
Financial pressures affect local authority marketing

As a result of the local authority's marketing strategy and conflict between its economic development objectives and wider financial constraints, the future of this small site of 0.5 hectares, available since 1985, remained uncertain in 1990.

The site was previously part of a British Rail depot and had lain derelict for five years before it was purchased by Wirral Metropolitan Borough Council in 1984 and reclaimed with the help of a derelict land grant of £47 000. The site was levelled and grassed in 1985 and was then available for development.

No development interest was expressed between 1985 and 1987, reflecting weak market demand. In 1987, the site was offered to an...

...adjacent shop fitting company at a price in the region of £45 000. The offer appears to have been treated by this company as an indefinite option and no reply was made to it until 1988. By this time, however, the market had improved and the local authority felt the site was worth considerably more than first suggested. A nearby food manufacturer, employing 150 people, expressed interest in the site and argued that without it, it would have to close its Seacombe factory and consolidate operations at Warrington. The local authority then decided to invite tenders for the site as a result of which, it was agreed in October 1989, to sell the freehold of the site to the food manufacturer. In 1990, before contracts were signed, the food manufacturer decided in any event, to close the Seacombe factory and the local authority therefore withdrew from the agreement.

Although the site was not formally remarketed, it remained available and interest existed. The future of the site, however, continued to be linked to that of the food processing factory. If the factory were ever to be subdivided into separate units, this site would be necessary to improve poor access and provide additional parking. The local authority would in the past have been happy to make the site available for this purpose, in order to protect existing levels of employment.

Instead the local authority had been impelled to maximize the value of its assets as a result of financial pressures. The site was well located for a petrol filling station and such a use would create a much higher land value than industrial development. Although it would create only three or four jobs, the local authority obtained planning permission for a garage and petrol filling station on the site in order to be able to market it for that purpose. In 1990, the future of the site remained uncertain.

Such behaviour should not be regarded as a valuation constraint but rather as an attempt to move land suitable for industrial development from that market to a more lucrative market sector. If potential industrial land can be sold for offices, housing or retailing, a much higher price can be achieved. Such attempts were made increasingly possible by less restrictive planning controls in the 1980s, including the introduction of the B1 use class in 1987. Competition from business uses subsequently contributed to the rapid rise in the price of good quality, well located industrial sites in the Cheshire—Wirral corridor in 1988 and 1989. The impact of such shock waves were reduced in time, however, as the market adjusted to new circumstances.

9.3.3 Passive land ownership

Central government policies to overcome passive ownership have focussed on the land held by public bodies. The principal statutory interests identified in the Cheshire—Wirral corridor have been local authorities and development agencies such as English Estates and the Merseyside Development Corporation. In line with their economic development objectives, many examples have been given of these ownership interests actively supplying sites. The scale of their activity ranges from the creation of the first phase of Crewe Business Park, where Crewe & Nantwich Borough Council together with Cheshire County Council made 35 hectares available (Case Study 100), to the provision of small plots which enable the expansion of inner city companies on constrained sites (Case Study 101).

Case Study 100
Promotion of business park by statutory landowners

In the late 1980s, Crewe & Nantwich Borough Council and Cheshire County Council successfully combined to make approximately 35 hectares of greenfield land available for private speculative development in the first phase of Crewe Business Park. The borough council originally owned a greater proportion of the site than the county council, but in order to equalize their respective contributions, the county council paid for a higher proportion of the servicing and infrastructure costs. Members and officers of both authorities worked closely on the entire project from design to marketing. A special steering committee was established and certain powers delegated to senior officers in order to minimize delays. Roads, services and other infrastructure for Phase 1 were completed by 1988 and plots were then marketed for business development by owner-occupiers or speculative developers. By 1990, approximately 11 hectares on the frontage of the business park had been acquired on 125 year ground leases for office rather than for manufacturing development.

Case Study 101
Local authority as land supplier

This case study provides a good example of the role of the public sector in enabling industrial development to occur in run down urban areas.

Since 1966, Robinsons had occupied a small site of 0.2 hectares close to Birkenhead docks, where the company manufactured marine refrigeration equipment. By 1985 the existing site and premises owned by the company had reached capacity. An adjacent site had been acquired by the local authority under an earlier slum clearance order. Although Wirral Metropolitan Borough Council was prepared to make the whole of this land available, Robinsons took only sufficient land for their immediate expansion. Industrial units were subsequently developed by Wirral MBC on part of the remaining land.

Regional assistance from the Department of Trade and Industry (£31 000) and a grant of £8000 for security fencing from Wirral MBC under the urban programme enabled Robinsons to proceed with the construction of a new light engineering unit. At the time, expansion of the business would have been considered a 'luxury' without this financial assistance from the public sector. The business continued to prosper and a further ten to 15 jobs were created as a result of the development of the new unit. By 1990, however, further expansion was again halted by the restricted size of the site. The company was able to buy a small plot of land from an adjacent scrapyard and purchase the remaining vacant part of the former housing clearance area from the local authority. Land uses within the overall site area were reorganized to provide a new car park, the construction of new offices and the conversion of the existing office block to industrial floorspace.

According to the company, this would be sufficient for the next ten to 15 years but thereafter future expansion could only occur if the company were to buy out the whole of the adjacent scrapyard. Although the company was no longer dependent upon the docks to transport its goods, it was reliant upon good motorway connections and access to both Liverpool and Manchester Airport. If it were ever to move, it would look first to remain within the local area. It would thus retain existing labour and be able to take advantage of any grants...

...from the Merseyside Development Corporation. Relocation outside the borough would only be considered in the last resort, if no appropriate site within the inner part of Wirral were available.

In fact, of those sites which were undeveloped at the end of 1989, 11 were identified where the owners were retaining them without any plans for their development or marketing in the foreseeable future (Table 9.2). In nine of these 11 cases, the owners were private production interests, and the behaviour of private landowners was often in marked contrast to that of statutory ownership interests. Of the 16 undeveloped sites owned by the local authorities at the end of 1989, ten were either already under offer or being actively marketed and there were plans to market or develop the remaining six. The multiple ownership constraints which affected the site of the possible extension to Crewe Business Park perhaps provide a more vivid picture than statistics alone, of the contrasting behaviour of private and statutory ownership interests (Case Study 102).

Out of the 11 undeveloped sites which their owners intended to retain without any plans for development, corporate inertia was the most single important constraint upon development in eight cases. Where there are no holding costs attached to retaining land and where the site comprises only a very small part of the ownership interests' total landholdings, there is little incentive or inclination to review the land management strategy with reference to these sites. Indeed in one example the site had been inadvertently omitted altogether from the property manager's register of potential sites (Case Study 103). Such corporate inertia becomes a matter for legitimate public concern when the aggregate supply of land is limited either by the planning system or simply because an area is already heavily built up.

Case Study 102
Ownership constraints on business park extension

Until 1990, this 8 hectare site on the north-eastern edge of Crewe, was let on an agricultural tenancy. The site is identified in the Crewe Local Plan for an extension of the nearby business park. Indeed in April 1989, the local planning authority obtained outline planning permission for B1 development covering this site...

...Two particular constraints, however, would have to be resolved before development could proceed. In the first place, the site was owned by the Crown Estate, whose normal policy is not to sell land freehold. Although the land has never been actively marketed, the Crown Estate would seek a financial package which offered a stream of rising ground rental income, under a long term ground lease. In the second place, access to the site is problematic. A direct connection to the business park would involve crossing land owned by a third party. This owner, who had already turned down one offer from the local authority, considerably in excess of his original purchase price, was believed to have bought the land purely for the purposes of speculation.

An alternative access directly onto the public highway in the other direction, which was previously considered possible, could be affected by accommodation works for a proposed bypass and development would not now be possible until the constraints identified above were resolved.

Case Study 103
Corporate inertia

This small site of 0.6 hectares, owned by British Telecom, was located to the rear of their premises in Upton, Wirral. Wirral Metropolitan Borough Council identified the site as suitable in principal for industrial development but recognized that physical and ownership constraints made development extremely unlikely other than by British Telecom or an adjacent company. In fact, the existence of the site has been largely ignored by British Telecom who, in the past, has omitted to include it in internal returns relating to their overall landholdings.

The whole site was first acquired by the General Post Office from the Ministry of Defence in 1974 and subsequently passed to British Telecom. There was no access to this area from the public highway except through British Telecom's operational area or by one of the adjoining companies. The site had previously been used to tip rubble and was now fenced off.

British Telecom has been prepared to lease another part of their operational area from the local authority, while allowing this freehold site to fall into neglect. No serious assessment of the site's potential had been made since 1986. A feasibility study commissioned by...

...British Telecom at that time suggested that the site could be used for the relocation of part of their operation, thereby releasing frontage land for sale. However, since the value of the frontage land, assessed at £160 000 was only slightly higher than the cost of bringing the landlocked site into operational use, at £153 000, no further action was taken. Furthermore, although an adjacent company had expressed interest in acquiring the site on three separate occasions, negotiations had never seriously been pursued. Following the last enquiry in 1987, the adjacent company was involved in a complicated land exchange with another neighbouring company in order to satisfy its requirements.

No pressure existed on British Telecom to market or develop this site and it was therefore likely to remain in its current state for the foreseeable future.

A number of production interests have been identified in the Cheshire—Wirral corridor whose fortunes declined during this century, but who have been slow to re-appraise their existing landholdings and their future operational needs. This has particularly applied to traditional industries such as ship building and also to the activities of the Mersey Docks & Harbour Company and the Manchester Ship Canal Company, discussed in Chapter 6. The latter two companies have not only been reluctant to release land in response to market demand (Case Study 104) but have again been prepared to do so only on terms and conditions which are favourable to their wider objectives (Case Study 105).

Case Study 104
Retention of land for operational use

This vacant cleared site of almost 8 hectares within Birkenhead Docks was owned by the Mersey Docks & Harbour Company (MDHC). Until recently, half the site was occupied by a flour mill, which was demolished when the mill operator surrendered the ground lease. At that time, MDHC created the other half of the site by filling two docks and an adjoining creek. Since there is a serviceable deep berth nearby, MDHC is only prepared to consider using the site for port related activities, whether or not they involve new building...

...No encouragement was therefore given by MDHC to approaches received at the height of the property boom from developers wishing to include this site in a major waterfront scheme. Similarly, proposals made by the local authority to use part of the site to relocate small users likely to be displaced from other land sold by MDHC to the Merseyside Development Corporation were not pursued owing to the disproportionate expense of a long internal access road and the possibility of objections from local residents.

In the meantime, an upturn in port activity in the late 1980s reinforced MDHC's view that the site should be retained for operational use.

Case Study 105
Meeting the specific requirements of the vendor

In 1966, an oil refining company established production on a 10 hectare site in Ellesmere Port on the banks of the ship canal. The land was acquired from the Manchester Ship Canal Company (MSCC) on a long term ground lease.

By the late 1980s, when the oil company drew up plans to double production capacity, it became necessary to acquire additional land. The company approached MSCC and was able to purchase a further 5 hectares in 1988, on a ground lease concurrent in length to that granted in 1966. The company relocated a lorry park for 150 vehicles from the original site onto the newly acquired land, allowing the proposed expansion to take place on the original site.

MSCC is usually reluctant to release surplus land through the open market. In this case, however, by enabling the oil company to increase production, MSCC achieved increased traffic on the waterways and hence increased business for itself through the disposal of its surplus land.

As well as the eight sites where development was inhibited by corporate inertia, passive ownership was also responsible for delays in marketing or developing some of the other 26 undeveloped sites, previously identified (Table 9.2) as being retained, at least in the short term, by their current owners. Passive ownership therefore extends from the superficial to the most intractable. The most intractable may prevent development altogether while the superficial will serve only to delay.

Passive ownership, in whatever form, clearly gives rise to conflict between suppliers and consumers. Conflict can be particularly intense where land needs to be transferred from 'declining' manufacturing and other production sectors to 'growing' sectors but this fails to materialize until the declining sectors appreciate the permanent nature of much of the restructuring of recent years. Passive land ownership thus reduces the supply of land offered to the market and undermines the quantity allocated for development by planning authorities. Unless land is already owned by land consumers, or is likely to be transferred without too much difficulty, it is not genuinely available. Aggregate land statistics which do not reflect the market availability of land for industrial development are of very limited value. Cheshire County Council has recognized this by identifying 'owner-specific land' separately in its published land availability statistics. The public sector, however, can play a direct role in overcoming passive land ownership, as will be discussed in the following section.

9.4 CHALLENGING PASSIVE OWNERSHIP

It is clear from the actual transactions identified and the ownership of sites still undeveloped at the end of 1989, that the key players in terms of land supply in the Cheshire—Wirral corridor were production interests and the statutory sector, particularly the local authorities. The ownership behaviour of some production interests, however, presented the most intractable supply constraints. Although ten undeveloped sites owned by major manufacturing interests were held off the market at the end of 1989, it can be difficult to assess the extent to which long term land retention by major manufacturing interests amounts to passive behaviour rather than to a rational and justifiable land management and development strategy. Indeed, past experience would suggest that, in some cases, investment and expansion will be undertaken by the company itself and land long kept idle in anticipation of development will be brought into use.

Where this does not arise, a number of catalysts were identified, particularly in Chapter 6, which may transform passive ownership behaviour and result in the voluntary release of potential development sites. It will be helpful to summarize these here before focussing on compulsory purchase procedures.

9.4.1 Reappraisal of future operational land requirements

It has been shown that a number of factors may lead companies to review the extent of their landholdings. Chief amongst these have been corporate restructuring; the impact of a rising property market and the eventual acceptance of the long term impact of industrial and economic restructuring. In a number of cases, companies came to the view that vacant land was unlikely ever to be brought into use for their own purposes and to the realization that it might command a high exchange value if offered in the open market (Case Study 106 and 107). Some major manufacturing companies were in fact undertaking feasibility studies in order to identify surplus land and prepare it for future marketing. Similarly, other production interests in transport and distribution had experienced a sharp decline in their demand for land. They had tended to retain under used sites, with development potential, for their own operational purposes, but some signs of greater realism were evident even amongst this group in response to potential market opportunities (Case Study 108).

Case Studies 106 and 107
Industrial restructuring and the release of surplus land

As a result of the decline of the shipbuilding industry, large areas of Cammell Laird's 50 hectare site in Wirral, which it had owned for over 100 years, became under utilized. The company reviewed its future land needs after privatization in 1987, receiving particular encouragement to do so from the Merseyside Development Corporation (MDC) in whose designated area the shipyard fell. In 1990, prior to the yard closing altogether, the company intended to contract its operational area, releasing land for development to the north-west, north, south and west of it. These four areas totalled 9 hectares. Of these, Case Study 106 comprised 1.9 hectares to the north-west and Case Study 107, 4.5 hectares to the south.

Site 106 had been cleared and was used in the 1980s as a marshalling area. Realization of the site's development potential in 1990 was delayed by physical constraints. It was an awkward shape, a problem which was to be resolved by the compulsory purchase of an area of land by MDC, to 'square off' the site. A more fundamental problem, however, was poor access to the public highway. This was...

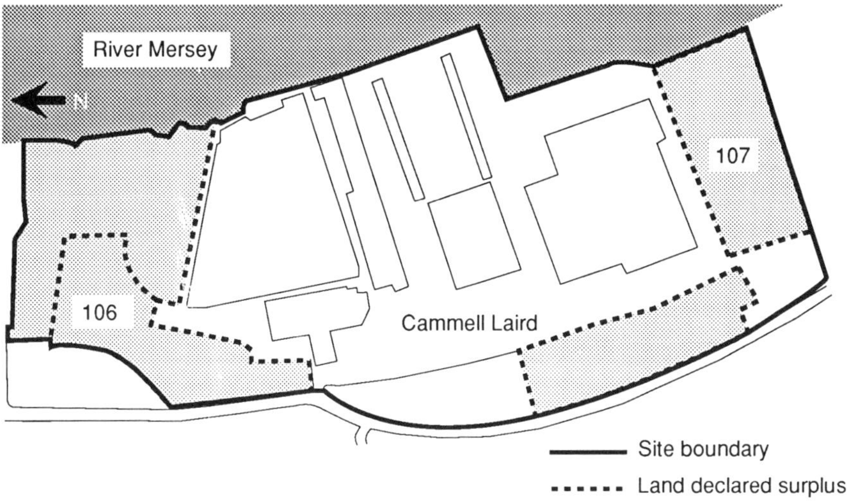

Fig. 9.4 Surplus land owned by Cammell Laird

...to be overcome by improvements to the A41 to be undertaken in 1991/92, a scheme funded by the Department of the Environment, the Department of Transport and MDC. It was anticipated that the plot would probably be developed for B1/B2 use.

In 1990 Site 107 was also a cleared area, let on licence for open storage. It had not yet been actively marketed but there had been interest from developers. It suffered from a number of physical constraints. Ground conditions were difficult, access to the public highway was poor and there were no internal site roads. In addition its proximity to a neighbouring oil terminal could strongly influence the extent and type of development interest.

Case Study 108
Decline in canal traffic and the release of surplus land

This 22 hectare site in Ellesmere Port was considered by the local authority to be well placed for employment development since it was adjacent to an existing industrial estate. It had been owned freehold since the 19th century by the Manchester Ship Canal Company in connection with its waterway operation. For at least 20 years,

...however, it had been largely unused, with the exception of 6 hectares to the south, which was leased for the commercial open storage of vehicles. As there had been no physical development of the site, it required servicing, surface drainage and improved internal access, together with improved access to the public highway.

In 1986, planning permission was given for retail development on the 6 hectares at the northern end of the site. This area had the most immediate access to the public highway. The company anticipated that the return on this development could be set against the considerable cost of bringing forward the backland. The retail proposal was not implemented, however, because of a lack of interest from major retail outlets.

From the end of 1989, that part of the site not covered by the retail permission was marketed for industrial use, in order to 'test the market'. While the company would consider offers for the land on the basis of a 125 year ground lease, it would prefer to develop the units itself having first secured prelets, and so benefit from the long term development gain. Although there was some interest in 1990, no agreements were concluded. It would appear that the rents considered necessary by the company, to make development economically viable, could not be achieved in the prevailing market.

The site was just one part of the company's surplus landholdings, and was accorded only a low priority. Although MSCC has become active in developing under utilized land for noncanal related uses, it has focussed on sites in Manchester where a higher return can be achieved.

9.4.2 Methods of acquisition by development interests

Land may well be acquired for industrial development without ever having been widely advertised or known to be available (Table 9.3). In the case of the 32 transactions involving sites now developed, acquisition by direct personal contact was almost as important as bargaining in the market place. Although the numbers are small, it is interesting to note that minor manufacturing companies appeared to rely much more heavily upon the open market than major manufacturing companies. It will be possible for passive behaviour by landowners to be challenged by direct approaches from

Table 9.3 Transactions in developed sites: method of acquisition

Ownership interest	*On market or known to be available*	*Acquired by by direct contact*	*Total*
1. Former landed property	-	-	-
2. Production, major manufacturing	2	4	6
3. Production, minor manufacturing	9	5	14
4. Production, other	-	1	1
5. Financial, investment	-	-	-
6. Financial, development	4	2	6
7. Statutory, local authority	-	-	-
8. Statutory, development agency	2	3	5
9. Statutory, other	-	-	-
TOTALS	17	15	32

development interests, which may again cause landowners to review their strategies and the opportunity costs of holding under utilized land. One circumstance in which direct approaches may be particularly likely to arise, as discussed in Chapter 6, is where the specific requirements of the purchaser tightly define and constrain the submarket in which it will be possible to identify a suitable site (Case Studies 109 and 110). It is reasonable to infer from the analysis of transactions that some of the undeveloped sites not yet on the market or even intended for retention, may well change hands in the future through personal and direct contact.

Case Study 109
Site acquisition by direct contact

This site, in a rural area close to Nantwich, was acquired by a chicken processing company. Unusually for a minor manufacturing interest, it was purchased, not through the formal network of agents and site marketing, but through informal and personal contact.

The 4.5 hectare site, which includes a farmhouse and outbuildings, was previously used as a chicken slaughterhouse. It was acquired in...

...1989 by its current owner, whose previous premises in Birmingham were subject to a compulsory purchase order. At that time, the slaughterhouse was fully operational and the site was not on the market. The site was known to the chicken processing company through trade links and an offer was made for it. The site was acquired freehold for £550 000 within 14 days of the first formal expression of interest. This mutually acceptable price was agreed between the parties concerned without recourse to professional advice.

It is evident that this particular site offered a number of advantages to the new owner. First, being in a similar use, it could be acquired very quickly without the precaution of applying for planning permission. Secondly, its remote location in a rural area was well suited to this unpleasant use, which would cause less offence or nuisance here than it would in a built up area. Finally, the size of the site offered the possibility of future expansion. The existing buildings were operational but in poor condition and a replacement building of around 1800 square metres was completed in November 1990, with plans also having been submitted for a further new building. The existing buildings were expected to be demolished, leaving room for further expansion.

Case Study 110
Takeover of company with suitable property

The company which owned this 0.8 hectare site on an established industrial estate in Chester was originally based in Letchworth. Its relocation in 1985 arose from a decision by the US parent company to expand its UK operation. It decided to achieve this by buying a similar industrial operation. The present site met its particular requirements. It provided two existing buildings, which could house two production processes separately, and afforded room for future expansion. Indeed, in 1990 the construction of a third building was completed, providing 600 square metres of new floorspace, and it was hoped to extend one of the older buildings within the following five years.

When the current owner acquired the site, it was not on the open market. Acquisition took nine months to complete. The purchaser paid £300 000 to acquire the freehold, a sum above the open market valuation.

9.4.3 The role of the public sector

Passive behaviour by ownership interests is exacerbated where sites experience severe physical constraints, whether on site or in relation to the surrounding infrastructure. The cost of overcoming difficult site conditions and problems involved in tackling off site constraints, such as inadequate highways, may be such that there is little incentive for landowners to actively market their sites and little financial advantage in undertaking remedial action.

The role of the public sector, whether in direct action or in providing financial assistance, can unlock these sites and transform passive behaviour. In Chapter 3, the role of English Estates was evident in conducting lengthy and complex negotiations with a recalcitrant owner (Case Study 33). An example was also given in the previous section of action by the MDC which would make surplus land long retained by Cammell Laird more marketable, using compulsory purchase procedures to resolve the problem of the site being an awkward shape and tackling necessary highway improvements together with other government departments (Case Study 106). Similarly, the availability of financial assistance may enable passive landowners to become active suppliers (Case Study 111) or even transform their role to that of developer. Merseyside Development Corporation, for example, was involved in a complex partnership arrangement under which grant aid was provided to the Mersey Docks and Harbour Company and Mowlem Plc to enable a prominent site in Birkenhead, long retained by MDHC for operational use, to be developed for business and industrial units (Case Study 112).

Case Study 111
Development Corporation finance for site preparation

This case study illustrates how the selective use of public funds can substantially improve the marketability of unsightly or derelict land. The site of approximately 1 hectare was located in Birkenhead docks and owned by the Mersey Docks & Harbour Company (MDHC). Until the 1970s, the site had been leased to British Rail for a shunting yard. When this use ceased, the site reverted to MDHC and was subsequently let, on a temporary basis, to a variety of small users, generating some rental income. By the time the last of these temporary users departed in 1988, the two buildings remaining from the British Rail days, had become extremely dilapidated and the remainder of...

...the site had attracted extensive fly tipping. MDHC's attempt to market the site at the time was confined to the erection of a site board, which, in view of the poor state of the site and the company's preference for port related users only, displayed a distinct lack of seriousness.

At the beginning of 1989, Merseyside Development Corporation (MDC) provided finance to cover 60% of the costs of site clearance and enhancement. The dilapidated buildings were demolished, fly tipping removed, perimeter fencing erected and the frontage landscaped. A total grant of £30 000 was provided for this purpose. By the end of 1989, MDHC had let the site, on 60 year ground leases, for development by three of their existing tenants from elsewhere within the dock estate.

Case Study 112
Redevelopment enabled by a complex development partnership

Mersey Docks & Harbour Company (MDHC) owns extensive areas of land in and around Birkenhead docks. Since the company derives the greater part of its income from berthing charges, port duties and so on, it normally seeks to retain land in operational use, often for such purposes as open storage and cargo handling. In fact, until recently, no financial advantage accrued from the freehold sale of land, since the proceeds were required to be distributed to shareholders rather than reinvested in the main business. The development of this site, in a manner quite exceptional for MDHC, demonstrates how financial assistance enabled an under used site to be developed and released a further 40 hectares of retained land to the MDC.

This prominent site of 3 hectares overlooking the River Mersey in the centre of Birkenhead was well located for bus, ferry and underground users. Until 20 years ago, the site had been used for the holding and slaughter of cattle imported from Ireland. When this use ceased, the premises were leased to a variety of small companies, including those in the engineering sector. The premises afforded a cheap, convenient and popular location for small companies, and at one time provided a rental income of £75 000 for MDHC. If the premises had not deteriorated and certain structures become dangerous, MDHC...

...would have been content to allow the occupation of the site on a low rent basis to continue.

In 1986, however, MDHC was approached by three separate developers interested in the site. By the end of that year, MDHC had invited each of them to submit details of their proposals and had selected one of the developers as their prospective partner. It was clear, however, to MDHC and to its development partner, that without some form of government assistance, estimated at 33% of development costs, the scheme would not be viable.

At the time, MDHC was in detailed negotiations with the Merseyside Development Corporation (MDC), which wished to acquire approximately 40 hectares of MDHC's land elsewhere, in order to promote dockland regeneration. As part of this deal, MDC agreed to assist and participate in the development of the 3 hectares site in Birkenhead.

A complex legal arrangement was established whereby MDHC granted a 200 year head lease to MDC, with MDC granting a sublease to the separate development company, which MDHC and its development partner agreed should be established between them for at least five years, with each party having a 50% interest. MDC were then prepared to contribute 33% of the development costs, in the form of assistance very similar to urban development grant. The remainder of the development costs were financed through an institutional arrangement negotiated by MDHC's development partner.

The site was cleared during the summer of 1988 and development commenced in the autumn. Three phases, with a total floorspace of 10 000 square metres, were proposed. Phase 1, comprising both offices and industrial units proved extremely successful, with 80% of units let on 25 year leases by the end of 1989. Phase 2 was likely to consist of only office floorspace, while further feasibility studies were intended to determine the exact mix of Phase 3.

9.4.4 Compulsory purchase

Local authorities can, as a last resort, seek to overcome passive ownership through compulsory purchase. A complex and lengthy set of rules, however, structures the relations between the public and private sectors in compulsory purchase (Denyer-Green, 1989). Even when compulsory purchase orders succeed, the compensation payable (making full allowance for disturbance, etc.), can be unduly expensive. This tends to discourage local authorities from initiating compulsory purchase orders in the first place. Two of the local authorities in the Cheshire—Wirral corridor, Wirral Metropolitan Borough Council and Ellesmere Port & Neston Borough Council, were unwilling as a matter of principle to contemplate compulsory purchase action. Compulsory purchase orders were regarded by Wirral Metropolitan Borough Council as 'very heavy handed' and therefore to be avoided, while Ellesmere Port & Neston Borough Council considered that compulsory purchase action would be likely to upset the council's long established and solid relationship with the major employers in the borough, such as Shell and Vauxhall.

In Chester, the approach to compulsory purchase action was slightly more pragmatic. While compulsory purchase orders were considered very much a matter of last resort, the city council was prepared to sanction such action provided that the cost was fully justified and that the action did not undermine the prospects for public/private sector cooperation in employment development. Crewe & Nantwich Borough Council were found to be the local authority in the Cheshire—Wirral corridor with the most recent experience of using compulsory purchase to promote employment development. The borough council did not oppose compulsory purchase on principle but preferred to acquire land by negotiation since this method was usually much quicker and compulsory purchase action is not always straightforward (Case Study 113).

Case Study 113
Developer fights compulsory purchase

Crewe & Nantwich Borough Council had identified a 7 hectare greenfield site and an adjoining area of 5 hectares for employment use in the draft Nantwich Local Plan in 1988. This allocation was made in recognition of the needs of local employers since there had been no new land allocated for industrial development in Nantwich for many years. In 1988, the local authority obtained outline permission to service the whole area. These two sites, however, were in separate ownerships.

The 7 hectare site was owned by a private trust which had entered into a conditional contract to sell to a private developer via a land...

...exchange. The prospective developer applied for planning permission to develop it in accordance with the local authority brief. The one departure was with reference to the proposed access to the site. The access originally outlined by the local authority passed over the adjacent land. If the 7 hectares were to be developed separately, this would no longer be feasible. The adjacent 5 hectares were in private ownership. A second developer had an option on this site and hoped to build housing. In order to ensure the comprehensive development of the entire area for employment use, the local authority issued compulsory purchase orders. Furthermore, planning permission for the proposal submitted for the 7 hectare site was refused on highway grounds. The prospective developer of this site, and his local agent, pursued this issue through a local plan inquiry and also appealed against the refusal of planning permission. Following the local plan inquiry, the inspector ruled that only the 7 hectare site should be developed for employment use and the adjoining 5 hectares should not be developed. This has been accepted by the local authority. The prospective developer also won the planning appeal in August 1990. Acquisition of the site was completed later that year, by which time the agents reported that the majority of the plots had already been taken.

Merseyside Development Corporation demonstrated a greater willingness than any of the local authorities in the Cheshire—Wirral corridor to tackle ownership constraints through compulsory purchase action (Case Study 114) and to use these procedures more generally in order to assemble viable and attractive sites for development (Case Study 115).

Case Study 114
Development Corporation vests ownership of site

Although this site of 4.2 hectares had been potentially available for development since 1985, poor access, lack of services and attempts by the previous statutory owner to secure planning permission for retail development impaired its market availability. Intervention by the development corporation released the site for employment use.

The site lay half a mile south of Birkenhead town centre. In 1982...

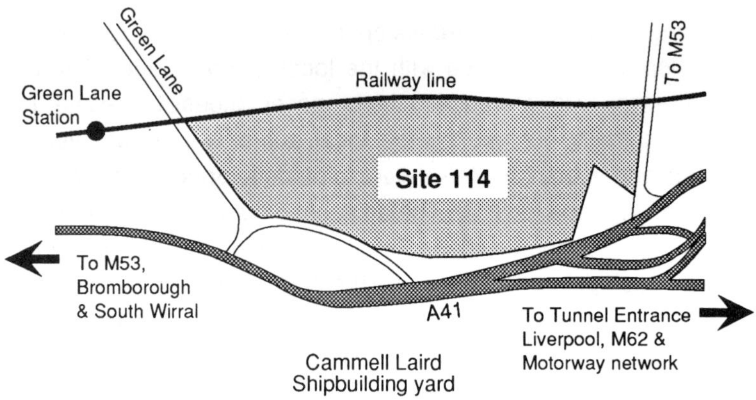

Fig. 9.5 Site acquired and marketed by MDC

...it was sold by British Rail to Merseyside County Council, who, with the assistance of derelict land grant, cleared it of high level railway sidings, and in 1985 obtained planning permission for a change of use to general industry.

With the abolition of the Metropolitan County Council in 1986, the site passed to the Merseyside Residuary Body, which had responsibility for disposing of the county council's assets for maximum value. Planning approval was therefore sought in 1986 for retail use. Wirral Metropolitan Borough Council, however, had identified two other sites suitable for retail development and sufficient to meet local needs. Permission was refused. An application in 1988, which was also turned down, went to appeal. The appeal was withdrawn when the Merseyside Development Corporation's designated area was extended to include this site. The Residuary Body then applied for retail permission from the development corporation, now the relevant planning authority. Again permission was refused, the MDC having adopted the local authority's shopping policy statement.

In July 1989, the development corporation vested the site in its ownership, together with an adjacent 0.5 hectares owned by the borough council and was concerned to bring it forward for development. Access to the site was poor. Improvements proposed by the MDC necessitated the acquisition of a number of small adjacent properties in order to give the site frontage onto the main A41. It was estimated...

...that the cost of these improvements and of bringing services on to the site would be in the region of £1 million. Improving access was seen as a necessary precondition of the disposal of the site, which was marketed in March 1990.

It was hoped that the site would be developed not only for employment generation light industry, offices and high technology use, but also for good quality 'front of house' activity such as motor showrooms and a leisure component.

Case Study 115
Development Corporation threatens compulsory purchase

In this example, the development corporation aimed to use its powers to enhance the area surrounding a prestigious development. The site in question within Birkenhead docks had been owned for the previous 15 years by an established company which operated a steel stockholding business with a £14 million turnover. It was in close proximity to the proposed Twelve Quays project, a waterfront development of approximately 500 houses and 60 000 square metres of business and commercial floorspace on 23 hectares.

Both the Twelve Quays development and this 1 hectare site fell within the designated area of the development corporation (MDC). To the MDC, the open storage of steel plates and the associated gantry and track equipment which occupied the site appeared incompatible with the environmental quality desired at Twelve Quays. Although the site was not included in the Twelve Quays planning application, it was covered by the developer's diagrammatic proposal 'for illustrative purposes'. In 1989, MDC therefore began the necessary procedures to acquire the 1 hectare site by compulsory purchase and the steel stockholding firm was advised accordingly.

When threatened by compulsory purchase, the company identified a better located site nearby with room for future expansion and entered into negotiations to take a long term lease. In the meantime, the property recession cast severe doubt over the implementation of the Twelve Quays scheme and the future use of the 1 hectare site was therefore extremely uncertain.

9.5 SUMMARY

Ownership constraints can only be viewed against a particular theoretical view of the processes of land supply and land exchange. Land and property markets do not function in a perfectly competitive manner. Willing buyers may not bring forward willing sellers. Only 12 undeveloped sites were being actively marketed or were known to be available at the beginning of 1990. Eleven further sites were under offer but over half the undeveloped sites were being kept off the market at that time. Some of the latter sites were intended to be marketed or developed by their owners in due course. However, a hard core of eight undeveloped sites (all but one owned by the private sector) were to be retained by their current owners, without any development proposed.

The case studies indicated that passive ownership has a potentially more serious impact on development potential than physical or infrastructural constraints, since it is less easy to pinpoint and more difficult to overcome. Furthermore, as the research in the Cheshire—Wirral corridor demonstrated, the public and political understanding of passive land ownership is often ill directed. Local authorities and government development agencies were generally very active players in the industrial land market. Apart from minor examples of the area health authority and the British Rail Property Board, passive behaviour in the Cheshire—Wirral corridor was almost exclusively confined to the private sector.

For philosophical reasons, it may be argued that passive ownership requires no intervention. Yet without more effective measures to tackle corporate inertia among declining production interests, a credibility gap will continue to exist between land allocation and land availability. In the final chapter, proposals are therefore made to facilitate making land available to industry, which include measures to tackle passive land ownership.

PART FOUR

CONCLUSIONS AND RECOMMENDATIONS

10

LAND FOR INDUSTRIAL DEVELOPMENT: TOWARDS A MORE EFFECTIVE POLICY RESPONSE

One essential component of British land policy in the 1980s was the emphasis placed upon reducing supply constraints on land development. In some cases, the frontiers of the state were rolled back as much greater flexibility was introduced into the town and country planning system and measures were taken to challenge passive land ownership in the public sector. In other cases, an enhanced role was accorded to the public sector. Much greater priority was given, for example, to preparing derelict sites for hard end uses, by tackling both physical and infrastructural constraints with grant aid. In certain locations, new institutional arrangements were established to spearhead the attack upon supply constraints. Urban development corporations, with their extensive powers and resources, became the best known example of this.

Many, but by no means all of these initiatives were directed towards the inner city. By the mid 1980s continuing inner urban decline was widely regarded as a powerful symbol of the limitations of macro economic supply side policies. After its re-election in 1987, the Conservative government devoted increased energy to addressing what it saw as the main problems of 'those inner cities' (Robson, 1988). The physical transformation of the inner urban environment achieved such prominence in British urban policy in the 1980s, that one commentator characterized such policy as stimulating urban regeneration through private sector property development (Healey, 1991b).

From a political perspective, attempts to promote development in the inner city have an added advantage. Those who protest most vehemently against the continued incursion of the development industry upon greenfield sites at the urban periphery often identify plentiful reserves of vacant and derelict land within existing urban areas, to which they believe development pressure ought to be directed. Ministers have sought to accommodate such views by vigorously encouraging the redevelopment of 'urban wasteland'. In reality, however, the limitations of such a policy linkage are all too apparent. Despite substantial public investment in improving the quality of urban land, private investment has focused on those locations where substantial development grants are on offer, or where flagship projects are underway, while development pressure has continued on peripheral greenfield and even green belt sites.

Earlier research (Adams, Baum and MacGregor, 1988) had suggested that in addition to physical constraints, ownership and valuation constraints were more likely to impede the availability of land for inner city development than planning restrictions. The urban land market is far from perfect, particularly when few transactions take place. Demand does not necessarily bring forward supply, since suppliers are not always profit maximizing in their behaviour. The imperfect nature of the land market also constrains locational choice for manufacturing industry.

Extensive urban land recycling programmes may well be able to tackle physical and infrastructural constraints to development. However, unless passive land ownership in both public and private sectors is tackled with similar determination, developable land will continue to be held off the market. Even if released, such land may well be valued by comparison with historic levels of value in the locality rather than on a realistic assessment of its potential. Complex institutional factors are thus as important in determining the speed of development as the severity of any particular physical or infrastructural constraint.

An agenda for research into the relationship between structure and agency in land and property development processes has been set out by Healey and Barrett (1990). While a rich variety of investigative work has been undertaken into the actions, interests and strategies of actors in the development process, much less attention has been directed to the ways in which broader processes frame or structure the behaviour of particular agents. Indeed, past theoretical approaches have lacked the capability to explore such fundamental interrelationships. According to Healey and Barrett, research which combines an understanding of both structure and agency is most likely to prove fruitful in the future. Such research would seek to explore ways in which agencies, individuals and organizations define and implement their strategies in relation to the resources available for development, the politico-juridicial rules which limit development opportunities and the ideas and values which people hold about the built form they would prefer and the environment they would seek. Such elements of structure are, however, neither permanent nor unchanging. Indeed, structure is established and re-established by the way agents* operate—by deploying, acknowledging, challenging, and potentially transforming rules and ideas as they frame and pursue their strategies. Agents react to the rules and imperatives of structural forces in different ways and, in doing so, help to reinforce or push at the edges of structure.

Such an approach helps us to set the impact of particular site constraints in a clearer context. While remedial action which tackles problematic site conditions may make development possible, whether in fact it brings development nearer will depend on the interests and strategies of relevant agents in relation to broader structural processes. Particular agents may be more active than others. The land management and development strategies of certain landowners may be more likely, for instance, to initiate site development than those of others

Research in the Cheshire—Wirral corridor therefore sought to move beyond a statistical analysis of the nature and distribution of development constraints and instead to concentrate upon the processes through which they are most likely to be overcome. Those agents which have succeeded in tackling problematic sites have been identified. After a brief review of the constraints identified in the research, this chapter will examine the relationship between structure and agency in industrial land development, before moving on to suggest a more effective policy response for the promotion of industrial development.

10.1 THE PATTERN OF DEVELOPMENT AND IMPACT OF SUPPLY CONSTRAINTS

Approximately 164 hectares in the Cheshire—Wirral corridor were either developed for manufacturing industry between 1985 and 1989 or were under development at the end of 1989. Almost 80% of the land taken up was located in major urban areas, in either Birkenhead or Wallasey (outside the official inner urban area) or in other settlements with a population of over 35 000 (Chester, Crewe and Ellesmere Port). The remaining 20% of the land taken up was split roughly evenly between inner urban, minor urban and rural areas. The speed of development increased quite markedly in the later years, reflecting the economic upturn among the sample sites. Among the sample sites, for instance, while development of 10.1 hectares was completed in 1987, 19.0 hectares were completed in 1989.

*The use of the term 'agents' in this chapter refers to all those who participate in the land and property development process. It is therefore used in a much wider sense than the earlier reference to property or estate agents, from Chapter 2 onwards.

Potential supply at the beginning of 1990 stood at 708 hectares, on 193 separate sites. At mean annual take up rates between 1986 and 1989, the potential supply of industrial land in the Cheshire—Wirral corridor would last for a further 22 years. This aggregate figure is, however, extremely misleading. On the one hand, it takes no account of market segmentation by size and location. On the other hand, inclusion within potential supply did not indicate that a site was available to purchase or was entirely free from physical constraints. Indeed, the quantity of potential industrial land gave very little indication of its likely quality or its suitability for immediate development. Detailed information was gathered on planning, physical, ownership and valuation constraints for 120 sample sites. Comparisons were drawn between the 60 sites which had been developed or were under construction and the 60 sites which remained undeveloped.

Remarkably little evidence was found that planning restraint constrained supply in any part of the Cheshire—Wirral corridor, apart from the particular case of Chester. Local planning policies evolved in order both to regenerate the urban economy and to diversify the rural economy. On only one sample site was industrial development prevented by the refusal of planning permission. At Chester, where demand was strongest at the time of the research, the city council had taken the lead to secure the release of 57 hectares from the green belt for new employment uses. A fierce campaign against this was mounted by conservation groups, who subsequently gained the support of the Secretary of State for the Environment. National policy objectives thus overruled local economic objectives.

Information was gathered on 13 different types of physical and infrastructural constraints. The existence of such constraints was discovered on 186 occasions, with a mean for sites already developed or under construction of 1.2 constraints per site and, for sites which remained undeveloped, of 1.9 constraints per site. The most widespread physical constraints were caused by poor access, redundant docks, tunnels and other structures, and by poor ground conditions or drainage. Apart from a greater preponderance of docks and structures in inner urban areas and a greater absence of infrastructural connections in rural areas, no noticeable variations existed in the spatial distribution of physical and infrastructural constraints.

While the existence of physical and infrastructural constraints was recorded on 186 occasions, on only 26% of those occasions did the constraint affect the development potential or marketability of the site. On developed sites such constraints had been overcome by the time of our interview and were therefore likely to be reduced in importance in the mind of the interviewee. Similarly, on undeveloped sites what may appear as a constraint, may in due course be eliminated and its impact minimized.

Ownership constraints can only be viewed against a particular theoretical view of the processes of land supply and land exchange. In the Cheshire—Wirral corridor, only 12 undeveloped sample sites were being actively marketed or were known to be available at the beginning of 1990. Eleven further sites were under offer but over half the undeveloped sample sites were being kept off the market at that time. Once prosperous industrial users, reluctant to release their historic landholdings, were particularly evident in the inner urban area. Case studies indicated that passive ownership was more difficult to tackle than physical or infrastructural constraints, which could be tackled by financial measures. Passive behaviour was concentrated amongst private sector ownership interests.

Previous work (MacGregor *et al.*, 1985) has suggested that even if owners are prepared to release land for development, they may take an unrealistic view of the likely value of the land. This may be considered as a valuation rather than an ownership constraint. Such valuation constraints however, appear to be of limited importance in the process of industrial development. Only three of the 120 sample industrial sites were affected by valuation dispute, of which only one was located in the inner city. The low incidence of valuation constraints may reflect the dominant role in the local land market of manufacturing interests rather than financial interests.

No simple relationship existed between planning, physical, ownership and valuation constraints and the speed with which industrial sites were brought forward for development in the Cheshire—Wirral corridor. Neither was there any strong pattern in the spatial distribution of constraints, which might explain urban—rural shifts in manufacturing employment. Poor quality sites were not necessarily left undeveloped while good quality sites were not necessarily developed. Too many academic and professional approaches tend to focus on the existence of constraints and appear to assume their static nature. The case studies showed that the impact of such constraints could be much reduced in an improving property market or by appropriate public sector action. Through ingenuity, resources or recourse to legal powers, most constraints could effectively be overcome. Indeed, tackling such constraints and making land available had become very much an accepted part of the process of development.

Most sites in the Cheshire—Wirral corridor were only developed for industrial purposes because the agencies, individuals and organizations involved had a fairly identifiable interest in overcoming site constraints and in taking the necessary action to bring land forward for development. On many sites which remained undeveloped, such action was already planned or was indeed underway by the end of 1989. The motivation and behaviour of landowners and developers, the activities of public sector agencies and organizations, and the impact of wider structural processes were of greater importance to the prospects of development than the mere existence of site constraints.

What is therefore required is a framework for analysis which takes a dynamic, rather than a static view of the impact of supply constraints and acknowledges the broader processes which set the context within which particular agents pursued their own strategies and interests. These processes affected the resources to which agents had access, the rules which governed their behaviour and the ideas which they drew upon in developing their strategy. The next three sections therefore describe and illustrate the interaction of agency behaviour in industrial land development with resources and the economy; with politico-juridicial rules and the state; and with cultural values and civil society. The final section brings these conclusions together as a basis for subsequent recommendations.

10.2 RESOURCES AND THE ECONOMY

Resources proved a critical influence upon the desire and the ability of agents to overcome unfavourable site conditions and bring land forward for development. Resources for private sector development were closely linked to market conditions. The demand for industrial land and premises rose sharply from mid 1988 to late 1989 in response to the economic upturn. Industrial land prices in Chester, for example, more than doubled from £135 000 per hectare in late 1988 to £295 000 per hectare a year later. Despite years of economic decline, prices also increased substantially at Birkenhead, although not so quickly as at Chester. Such rapid increases were attributable to the almost simultaneous desire of many manufacturing companies to expand, after several years in which little new construction had taken place. Such pressure for immediate development often provided the resources necessary to tackle longstanding site difficulties. In some cases, companies under pressure to expand onto adjacent land were willing to meet the necessary expense of bringing that land forward for development. In other cases, the upward trend in rental levels overturned earlier negative land valuations, enabling local developers to tackle difficult sites profitably.

Most of the Cheshire—Wirral corridor could not, however, be considered a 'prime location', according to the criteria acknowledged by most speculative developers and investors. As a result the development boom of the late 1980s encouraged very little private speculative development of industrial floorspace in the corridor, in comparison with the floorspace constructed by manufacturing interests themselves or by the public sector. Such private speculative activity tended to be by local or regional development interests and concentrated on refurbishment. National development companies would argue that the growth in industrial rents was neither sufficiently rapid nor permanent to justify extensive new construction. Capital flows are also likely to be attracted not only by locations where market returns are highest, but by those where they are most certain.

Since information within markets passes by word of mouth between established players, it is extremely difficult for possible future operators to know when best to break into the market. The informal and decentralized nature of land and property markets thus have important structural implications for the resources devoted to speculative industrial development in particular locations. More formal mechanisms for the exchange of market information may help to generate confidence and attract an inflow of speculative development capital.

In contrast, public sector resources for industrial development respond not to market conditions but reflect political realities. Capital expenditure by local authorities has been increasingly squeezed in recent years. One case study demonstrated how a small development of two 500 square metre advance units was recently undertaken in inner Birkenhead by Wirral Metropolitan Borough Council (Case Study 5). Although the cost of the development was 75% funded through the urban programme, it had to be phased over two financial years in order that the local authority's own contribution could be covered. On a nearby site, again owned by the local authority, an offer of £140 000 was received from a manufacturing company in October 1989. The company, however, withdrew before contracts were signed. Subsequently, the local authority obtained planning permission on the site for a petrol filling station. Although this latter proposal would create fewer jobs, it would maximize the value of the site at a time of severe financial difficulties for the authority.

Although the role of local authorities in promoting economic development has been acknowledged in legislation, central government constraints on capital allocations to local authorities have increasingly limited the role which the latter can play. The amount which local authorities can spend on capital projects, whether funded through grant aid or their ability to borrow money on the financial markets, has been cut dramatically in real terms in recent years. In addition, the phasing out of the urban programme has implications for areas such as Ellesmere Port and Wirral where the local authorities have been particularly active in constructing advance units, and have benefited significantly from this additional source of capital funding. The ending of the urban programme exacerbates the concern of these authorities about their future role and about the supply of industrial property in areas where the private development sector is not active. Central government's increasing emphasis on more targeted, and spatially limited initiatives such as 'city challenge' and 'city pride' does not address these problems.

Each of the local authorities within the Cheshire—Wirral corridor had formulated economic development strategies, as part of which they sought to increase the flow of resources into the area. Variety characterized the content of these strategies but each involved enhanced cooperation between public and private sectors. For example, in Crewe & Nantwich, the borough council formed a joint venture company with Bovis Urban Renewal Limited in order to bring together the authority's local capital and land resources with more significant development funding and expertise. Chester City Council, as part of its wider economic development strategy coordinated widespread lobbying at the national level to improve the city's rail links and hence its accessibility for investment. Ellesmere Port & Neston Borough Council organized an Annual Business Show and was also prepared to put time and resources into helping manufacturing companies make applications to the Department of Trade and Industry for regional assistance. Wirral Metropolitan Borough Council lobbied central government for more funds to be made available to encourage the private sector to undertake the reclamation and servicing of problematic industrial sites. Such activities again illustrate local determination to challenge and overturn unfavourable structural conditions.

Reassessment of spending priorities in the public sector not only affects local authorities. In a rural case study to the south of Chester, English Estates, acting on behalf of the Rural Development Commission, constructed ten small units in 1986/87 as the first phase of a village industrial estate. However, in 1988, the Rural Development Commission withdrew 100% funding for the second phase in order to concentrate its resources elsewhere and English Estates had subsequently to sell the site. In April 1990, the Secretary of State for Trade and Industry announced that English Estates itself would be required to reduce its own direct development role as a whole and begin to operate a new grant scheme to encourage private sector development.

English Estates is now absorbed into the new urban regeneration agency 'English Partnerships'. The role of English Partnerships is to promote the development of vacant, derelict and contaminated land, stimulate local enterprise, create jobs and improve the environment. The emphasis, however, will be on stimulating private sector involvement in areas where it has previously been weak and in fostering local partnerships between a host of agents interested in the development process. It will not assume the role which English Estates once performed. It is not primarily a development agency in its own right but is a funding partner which aims to encourage developers' confidence. It may undertake direct development but only if this *'will not disadvantage private sector development or investment in the area and* (that) *it cannot achieve its aim through other means'* (DoE, 1993). This study has shown the important roles played by English Estates, not only as a developer, but as a supplier of serviced sites. It will be some time before the impact of these changes on land supply in the assisted areas can be assessed.

Most public sector resources to support industrial development in the Cheshire—Wirral corridor come in the form of one type of grant or another. Assistance can be obtained by both public and private sector towards the costs of reclaiming and servicing industrial sites and, in certain cases, towards constructing industrial property itself. The variety of grants available appears confusing, other than to those with detailed and specialist knowledge. Sources include the European Regional Development Fund, derelict land grant, regional assistance and city grant. Grant availability often determined whether case study sites were developed or not. The complexities of applying for such grants, the uncertainties in securing them and in some cases the delay in actually making payment, combined to demand a certain persistence in those seeking to obtain such resources and to lengthen the development process (Case Study 116) .

A number of structural changes were announced after this study was completed which had the intention both of streamlining the process of applying for grant aid and of enabling a greater degree of flexibility and responsiveness to local needs than was previously possible in the administration of funding programmes. From April 1994 20 different funding programmes, including city challenge, regional enterprise grants and urban development corporations, were notionally brought together in the Single Regeneration Budget, which will become fully operational from April 1995.

With the exception of money earmarked for nondepartmental bodies such as English Partnerships, the budget is administered by the new combined regional offices of four government departments (Environment, Trade and Industry, Employment and Transport). These integrated offices bid for a share of the single budget and produce an annual regeneration statement, setting out priorities for regeneration and economic development. Although the detail of how the new system will operate is still evolving, it does mean that business and local authorities have only one port of call and there is the possibility within regions to shift resources between expenditure programmes. No additional resources have been made available under this regime. In 1994/95 the total budget was in the region of £1.4 billion. Of this English Partnerships share was to be between £250 million and £300 million.

Case Study 116
Developer decides not to seek grant aid

In 1989, a small local development company acquired a 1.3 hectare site in Wallasey which was almost entirely covered by an existing factory, dating from the beginning of this century. The total floorspace was almost 10 000 square metres. It had been largely vacant for seven years and had fallen into disrepair. This together with the size of the building had deterred interest in the site, particularly during a period of weak market demand. This developer, however, recognized the potential of the site in a relatively buoyant market and offered the asking price of £300 000, subject to completion within two weeks.

Work commenced in 1990 and completion was anticipated by May 1991. Some demolition was undertaken to provide parking space. The remaining 8000 square metres was to be converted into 29 industrial units and three office blocks, for sale on 999 year leases or to let on six year leases. By autumn 1990, around 25% of the space had been disposed of prior to completion. Despite the scale of the project, and the quality of the conversion and refurbishment which has in turn made it attractive to tenants, the developer has not sought grant aid, despite having been advised to apply by the local authority. The reason for this was his experience on an earlier scheme, where 18 months of what he described as "horsetrading" considerably delayed the project and resulted in a grant of £31 000 which barely covered the interest charges of £29 000 incurred in the meantime.

10.3 POLITICO-JURIDICIAL RULES AND THE STATE

Healey and Barrett (1990) argue that the state is unavoidably a critical factor in the development process. The state seeks not only to support the development process, but also to moderate its adverse externalities, safeguard social needs and conserve resources and environmental heritage. The rules and regulations of the planning system, which reflect a powerful set of values about what the environment should be like, have regulated land and property development. This has helped sustain land value growth and produce less risky and more protected markets than, for example, in the USA. Healey and Barrett point to the potentially destabilizing impact of deregulation and privatization upon the development industry in the UK.

The politico-juridicial rules of the state, which act as the second critical structural influence upon land and property development, are open to continuous interpretation and re-interpretation. Recent changes in the Town & Country Planning (Use Classes) Order have restricted the ability of local planning authorities to prevent office development on land previously allocated for industry. Another example in the Cheshire—Wirral corridor has been the evolution of green belt policy. In the past, the Merseyside and North Cheshire Green Belts, which restricted development in the western half of Wirral, in the Chester area and generally north of the A51 were rigidly enforced. More recently, more flexibility has been introduced into the operation of green belt policies. Redundant farm buildings have been allowed to be converted to enterprise workshops, to aid rural diversification. Elsewhere, insets have been more readily drawn around existing development in the green belt, in order to allow expansion and intensification to take place. On a more strategic basis, Chester City Council's attempt to release sites at the edge of the city from the green belt for employment development was over ridden by the Secretary of State for the Environment. However, in Chester, Crewe and in Ellesmere Port substantial greenfield sites on the periphery of the settlements, although not within the green belt, have recently been allocated for business park development. In Nantwich a peripheral greenfield site has been allocated for more general industrial development.

A further set of rules which structure the relations between the public and private sectors in land development concern compulsory purchase. Local authorities were deterred from taking compulsory purchase action to overcome passive ownership, not simply by financial considerations but also by what they regarded as excessively complicated legal procedures. The inadequacy of existing rules and procedures to overcome passive land ownership or prevent owners from allowing land to fall into an unsightly or derelict condition limited the market availability of land for industrial development in the Cheshire—Wirral corridor. Indeed, existing rules allow owners to abandon sites in a derelict condition, once use ceases, rather than devote expenditure to their restoration.

Since much of the industrial development which took place in the Cheshire—Wirral corridor between 1985 and 1989 depended upon one form of grant or another, the rules and regulations of the grant economy were an additional important structural influence. Although these rules and regulations were designed to achieve value for money, they were often regarded by those involved in the development process as obstacles to be carefully negotiated. Failure to do so could result in lengthy delays in, or even abandonment of, development projects. Case Studies 117 and 118 illustrate two cases where ingenuity helped overcome discriminatory rules and regulations which militated against reclamation by the private sector.

Case Studies 117 and 118
Ingenuity overcomes limitations of derelict land grant

Many of the case studies have shown the importance of grant aid to industrial development. In particular, derelict land grant was often critical to bringing sites forward. In these two cases, however, the lower proportion of costs available to the private sector resulted in the transfer of land to the public sector on a leasehold basis in order that it might be reclaimed with the benefit of 100% derelict land grant. Private landowners were eligible for only 80% of costs with no assistance towards infrastructure.

Both of the sites in question were large. Site 117 comprised just over 6 hectares in Ellesmere Port, owned by the Manchester Ship Canal Company on a 999 year lease. Existing warehouses were derelict, the site had been tipped, access was inadequate and in order to redevelop the site a filled in canal outlet needed to be excavated, properly filled and compacted. The site was sold to the local authority on a 939 year lease in the early 1980s. At this time the MSCC did not anticipate re-using the site but a pre-emption clause was included in the lease which effectively gave them first refusal when the site was subsequently offered for sale. The site was reclaimed and serviced by the local authority in 1984 and the bulk of it has indeed been reacquired by the MSCC at full market value, as port related activity had become more buoyant. The land is now used for warehousing, car parking and as an extension to the container port. English Estates, however, acquired 0.8 hectares from the local authority in 1985 and built six factory units of between 300 and 500 square metres...

Fig. 10.1 Part of MSCC site reclaimed by local authority and developed by English Estates (Case Study 117)

...Similarly Site 118 of just over 6 hectares in Bromborough, Wirral, which had been owned freehold for over 50 years by a major manufacturing company, was leased to Merseyside County Council on a 40 year lease. Again the lease included clauses to ensure that the freeholder could repurchase it on payment of any betterment arising as a result of the reclamation. Termination of the lease was in fact under negotiation in 1990. In the meantime the site had passed to the Merseyside Residuary Body, on abolition of the county council, and subsequently to Wirral Metropolitan Borough Council who reclaimed the site in the early 1980s, clearing it of a number of large concrete buildings. The cost of reclamation was £0.5 million. In 1990 negotiations were underway between the original freeholder and a drinks manufacturer who proposed to acquire the sale on a 150 year lease.

Finally, European legislation has become an increasingly important consideration in the development process in the UK, particularly in the sphere of environmental legislation. For example, the Urban Waste Water Directive (91/271/EEC) required all member states, including the UK, to introduce measures by mid 1993, to ensure that manufacturing companies have prior consent for any discharges of industrial waste in either controlled waters (rivers and lakes) or sewers. The directive was implemented in the UK through regulations under the Water Act 1989 and the Environmental Protection Act 1990. Such new regulations can constrain the locational choice particularly of heavy manufacturing industry, if either the National Rivers Authority (responsible for discharge consents to controlled waters in England and Wales) or Her Majesty's Inspectorate of Pollution (responsible for discharge consents of 'red list' substances to sewers) or the sewerage undertaker (responsible for discharge consents of other substances to sewers) appear likely to refuse consent at a specified site.

10.4 CULTURAL VALUES AND CIVIL SOCIETY

The ideas and values which agents draw upon in developing their strategies form an important element of structure. Contradictions are often inherent in such ideas and values and are only resolved on a temporary basis, as conflict between agents takes place from site to site or from year to year. Sometimes, certain agents appear to be gaining the upper hand. One useful example of such continuous conflict is the balance which is struck and restruck between amenity and development. To attract inward investment, local planning authorities often feel the need to sacrifice previously protected environments for new jobs. This reflects changing ideas and values about the quality of industrial development. Many years ago, local authorities were quite prepared to relegate industrial development to the forgotten corner of the town, out of site beyond the railway line. Industry could be characterized as the poor cousin amongst land uses. Protection of greenfield sites from major industrial development was subsequently felt to reinforce urban regeneration. Industry's land demands appear to be changing. Manufacturing companies are increasingly becoming image conscious and attach a value to prestige sites which may not reflect their actual development potential (Case Studies 119 and 120).

Those local authorities only able to offer poor quality sites may thus find themselves at a disadvantage. In the interests of jobs, previously protected environments have been released to provide high quality sites for employment development. Sceptics may wonder, however, whether such sites are needed to meet the production requirements of industry rather than the image requirements of its senior managers.

Case Study 119
An image conscious manufacturing company

An international manufacturing company which first established a small plant in Wirral in 1969 moved in 1978 to its present site in order to expand. This site lies on a small industrial estate to the north of Birkenhead docks and was previously cleared land which was left over from the construction of the new Mersey Tunnel approach road.

In 1986, the company also acquired an adjacent warehouse of approximately 600 square metres, which had become vacant when its previous owner went into liquidation. There remained room for further expansion between the original factory and the recently acquired warehouse, but whether any further investment would take place at this location would depend on the parent company. In fact, head office had expressed concern at the poor quality of the local environment. This was attributable to the extensive spillage of litter from the nearby local authority waste disposal site and particularly to the presence of 'bad neighbours' such as scrap metal dealers whom, when the company originally purchased the site, were expected to be relocated. Although the company was very keen to retain its existing workforce, environmental factors could persuade it to move out of the inner area.

Case Study 120
Landscape constraints on expansion in a rural area

This prestigious and unique site in a rural setting east of Crewe has been occupied since the mid 1960s by an internationally known pharmaceutical company. The site was leased from the Crown and comprises 18 hectares and an imposing Grade 1 listed building, dating from the 17th century, but substantially rebuilt and extended over the years. This house is set in 10 hectares of landscaped grounds and the maintenance of these grounds and the house is a condition of the 99 year lease.

Although the main building is used as administrative offices, the manufacturing and printing works are housed in other buildings, many of which have been purpose built since the 1970s. These modern buildings are totally screened and are unobtrusive. The size of the...

> ...site has enabled several new developments to take place in recent years. These included office extensions and loading bays, for example, and approximately 1000 people work on this site. Because the landscaped areas must be retained, and because of the sensitivity of any visible new building, there is now only limited scope for further new development. This constraint, however, would not induce the company to relocate — in part because of its existing investment here, but also because of the prestigious nature of this site. Future expansion can be accommodated at other large sites in the UK.

What is perceived to be the correct boundary between public sector action and private sector initiative is a further important component of the ideas and values which structure the industrial land development process. While vacant land owned by the public sector has to be publicly registered, there is no such requirement for vacant land owned by the private sector. The contrasting views and experience of compulsory purchase among the local authorities in the Cheshire—Wirral corridor reflects a wider debate on the extent to which the public sector should interfere in private property in order to promote industrial development. Indeed 'the private property ethic' has, on certain occasions, allowed passive ownership to prevent industrial development. However, more than once, the urgency of promoting urban regeneration required the Merseyside Development Corporation to challenge this ethic head on, and so overcome the particular ownership constraints.

Private property itself is subject to changing ideas and values. Whereas the 20—5 year institutional lease was a common form of tenure in the past, evidence now suggests that those industrial users who choose to rent are much more interested in taking shorter leases in order to retain flexibility. Companies which have climbed up the property ladder and envisage a longer term commitment at a particular location appear to show an increasing preference for obtaining freehold ownership. The research suggested that conflict over the proposed form of tenure between vendors and purchasers of potential industrial sites could delay development.

10.5 THEORETICAL AND METHODOLOGICAL CONCLUSIONS

A study of the industrial land market in the Cheshire—Wirral corridor showed that its operation was influenced by the interests, strategies and actions of agents in three particular ways. First, agents help define submarkets. For industrial land, the user submarket is, for instance, very different from the developer submarket. Manufacturing companies assess the value of land as users of industrial space and often have very different site requirements from developers, who concentrate on exchange value. Secondly, not all agents are profit maximizing as far as the use and development of land is concerned. Preferences of individual landowners may be irrational and corporate inertia is evident. Thirdly, few agents possess widespread knowledge and uncertainty is prevalent in the market. The research therefore confirmed, for the industrial land market, earlier work which had established the imperfect nature of urban land markets more generally.

In the urban land development process, the importance of development cycles was highlighted in many case studies. Timing is often crucial to overcoming physical constraints, for at the beginning of an upturn, high development costs may well be covered by rapidly rising rents. Passive ownership is of course much more difficult to overcome. Public sector action can play a critical role, however, in stimulating development. In several case studies, development was successfully promoted through a helpful package of public sector financial support and intervention.

The research demonstrated that the actions, interests and strategies of agents were heavily influenced by resources, rules, ideas and values. Such structural influences were, however, never constant but subject to a process of continuous change. Structure may at one moment favour industrial development but only a few months later may become entirely unfavourable if, for instance, an economic downturn leads to the withdrawal of resources. The proposed prestigious Twelve Quays scheme, a mix of housing and commercial development in Birkenhead, provided a good example of this. The concept was launched by Merseyside Development Corporation in October 1989 at the height of the property boom. The subsequent collapse of the property market has restricted progress towards implementation. New elements can also be introduced into the structural framework, which change perceptions as to what is and is not possible. For instance, the abolition of the National Dock Labour Scheme has encouraged the Mersey Docks & Harbour Company to take a more optimistic view of the prospects for port related developments. The company was consequently less willing to release under utilized operational land for other development purposes. Structure is therefore continuously on the move, and, as it interacts with the changing actions, interests and strategies of particular agents, development is either progressed or delayed.

Crucially, as Chapter 9 demonstrated, agents may play a variety of roles in the process of land supply, land exchange and land development. Active land suppliers may be more able to take advantage of favourable structural conditions and more willing to challenge and attempt to transform unfavourable structural conditions. Furthermore, resources, rules, ideas and values are not only reflected in the actions, interests and strategies of land consumers. Some consumers may in fact be more successful than others in benefiting from such structural influences and more willing to challenge structural conditions when they are to their disadvantage. The actions, interests and strategies of particular agents in relation to broader processes may vary from the benign to the malign. For example, some developers in the Cheshire—Wirral corridor took advantage of the improving property market in 1988—9, drew in resources, overcame problematic site conditions and constructed or refurbished industrial premises. Others, more influenced by their own cultural perception of 'prime' property and 'prime' locations, continued to boycott the corridor. The actions, interests and strategies of certain developers in relation to broader structural process could therefore be considered a benefit to development, while those of others, a hindrance.

10.6 RECOMMENDATIONS

Research in the Cheshire—Wirral corridor discovered no close relationship between the existence of site constraints and the likelihood of development. Most of the sites examined were, or had been, constrained in some way by physical or institutional factors. Good quality sites were not necessarily developed and poor quality sites were not necessarily left undeveloped. It must be concluded that too much attention is often given to the mere existence of constraints and not enough to the processes through which they are most likely to be overcome. Particular individuals and organizations are more successful than others in bringing problematic sites forward for development through the way in which their actions, interests and strategies are formulated and reformulated in relation to broader structural processes. A more effective policy response is needed to speed up the flow of land for industrial development in order to satisfy demand, particularly at boom times. Such a response would seek to avoid undue concentration on the existence of site constraints and would make no simplistic division between 'constrained' and 'unconstrained' sites. Instead, emphasis should be placed upon encouraging those individuals and organizations, in the private and public sectors, most likely to challenge and overcome site constraints. Equally it is important to confront those who are either least likely to do so, or are in fact, themselves the root cause of the particular constraints.

10.6.1 Information sources

In many areas, remarkably little information of any accuracy is published at a local level on property markets and development processes. This breeds uncertainty and leads to lack of confidence. A valuable service would be performed if all the existing knowledge on local markets and development could be brought together in a methodologically rigorous manner and published at regular intervals. This would involve close cooperation between local authorities, district valuers, regional and local property agents and others involved in the property and development business. Those involved in recent transactions, for example, would be able to contribute information about the immediate supply and demand for industrial land and property. A longer term perspective could be provided by local authorities, drawing on their planning and economic development records. In particular, local authorities should aim to monitor development activity alongside their existing records of planning applications, where they do not already do so. In some locations, local authorities may be the appropriate agency to publish comprehensive market information on a regular basis, while elsewhere this may be better undertaken by enterprise agencies, research institutes or other organizations. Accurate and reliable information at the local level would make a vital contribution to the smoother operation of market processes by, for example, pinpointing particular shortages at an early stage and providing the confidence for development to commence as soon as an upward trend in rental levels became apparent.

10.6.2 Market related planning policies

Better information would enable local authorities to devise policies which more closely relate to market processes and which seek to influence them more directly. More account could be taken of particular market sectors in the allocation of land. Some sites, for example, may be better suited to meet the requirements of manufacturing companies rather than those of speculative developers. Within the manufacturing sector, some sites may be sufficiently large and well located to be of regional significance, while others may only be of subregional or local importance. Where the supply of land is shown to be insufficient in particular market sectors in relation to disaggregated information on take up rates, then more might be made available.

10.6.3 Active land suppliers

The research showed that local authorities and statutory development agencies were at the forefront of reclaiming or servicing industrial land and making it available to users and developers. Local authorities, English Estates and development corporations have played a critical role in site preparation and land supply in anticipation of demand. Bottlenecks during periods of increased demand will clearly have implications for economic growth. A recession is certainly not the time to allow the allocation, preparation and servicing of industrial sites to slow down. Rather, the reverse is true. In order to catch the next development boom, local authorities and other statutory development agencies should continue to prepare land for industrial development. Every effort should be made to ensure that sufficient serviced land is available 'off the shelf' with the benefit of outline planning permission (for which local authorities could themselves apply), before demand starts to rise. However, care needs to be taken not to flood the market with a substantial over supply of potential development land since this may depress prospects for rental and capital growth and so frighten away developers and investors. The public sector should also continue to fill gaps in the property ladder through the development of advanced units to let. Those areas which are best prepared for the next development boom are most likely to benefit from it.

It is unlikely even in the new climate of partnership being encouraged between public and private sectors, that private development interests would be willing to undertake advance site preparation and construction in areas such as the north-west, while returns remain low and market confidence is weak. The statutory sector will continue to have a key role and there is a need for resources to support this activity. The distribution of spending powers within local authority capital allocations determined by central government must acknowledge local authorities' responsibility for economic development activity. It could be argued that this might be one way in which central government could best achieve its stated goal of achieving a funding regime which is more responsive to local objectives. It will be important to monitor the impact of the demise of English Estates and the reduced role of English

Partnerships in development activity in a climate in which local authorities continue to experience severe financial constraints.

10.6.4 Passive land suppliers

The research showed that the supply of land for industrial development was impeded by passive land ownership, particularly in the private sector. There is indeed little justification for maintaining the distinction between public and private sectors, so far as passive land ownership is concerned. Local authorities could collect much more information on the motives and behaviour of landowners. Public registers of vacant land should be extended to the private sector, as has been considered by the government. The long term retention of suitable development land by passive owners should be challenged, irrespective of whether the owner is in the public or private sectors. Where resources allow, local authorities could make greater use of existing compulsory purchase powers to secure the release of land held by passive owners in the private sector. More research, however, is needed to examine the effectiveness of existing measures to promote the transfer of land from passive to active ownership, including compulsory purchase procedures, public land registers and waste land notices. This work could usefully assess how the recently granted public access to H. M. Land Registry might contribute to overcoming passive land ownership.

10.6.5 Land consumers: manufacturing companies

Manufacturing interests have been shown to be a principal player in local land markets. Achieving their development objectives, however, often depended critically on the availability of financial assistance, whether regional assistance or derelict land grant. Such support must remain an important component of the single regeneration budget. Economic development in areas such as the north-west depends on a dual strategy which includes removing supply constraints but must also include stimulating and sustaining market demand. In particular, some form of derelict land grant, administered by English Partnerships has a continuing role to play and proposals for its reformation are given in Section 10.6.8. However, given English Partnerships' focus on urban priority and assisted areas, there must be a danger that other areas will be neglected. Despite the fact that English Partnerships is intended to continue the present work of derelict land grant in rural areas, limited resources and competing demands, such as the needs of communities once dependent on coal mining, may redefine priorities.

10.6.6 Land consumers: developers and investors

Developers and investors, who operated regionally or nationally, showed very little interest in the provision of speculative industrial accommodation in the Cheshire—Wirral corridor, apart perhaps from in Chester. Widespread and continuous activity by the development industry would probably require a sustained period of rental growth, combined with constraints upon the available supply of land. However, a small number of local experienced developers were identified who were able to put together successful schemes, mainly through a combination of refurbishment and new development. Such developers offer an invaluable local resource, which local authorities need to identify and encourage. Other research (Healey *et al.*, 1992) has shown how the implementation of urban policy in the 1980s, which engaged national companies in flagship projects, has been to the potential detriment of some locally based elements of the development industry (p. 287). The study of the Cheshire—Wirral corridor suggests that such local interests are well placed to gauge local markets and to respond quickly and flexibly to them.

Institutional investors could discover novel methods to invest in industrial property, in view of the growing unpopularity of the standard 20—5 year lease. For instance, where companies need to relocate to a new site, local authorities could work in partnership with institutional investors to make this happen. The local authority may be able to provide a site while the institution could offer a long term conventional mortgage. Alternatively, 'a displaced sale and lease back' could be arranged. The institution would acquire an existing industrial site for redevelopment, previously approved by the local planing authority, for which the company would realize a capital sum towards relocation. This might be converted into a rent free period on new premises constructed by the institution on a site provided by the local authority.

10.6.7 Industrial dereliction

Industrial decline in many parts of the country is continuously turning into industrial dereliction. As more and more industrial buildings come to the end of their useful life, new mechanisms are needed to ensure that sites are recycled rather than remain derelict. Those responsible for industrial dereliction should, in the long term, bear the cost of reclamation. In early 1992, the Department of the Environment issued a consultation paper and invited views on this matter (DoE, 1992b). Mineral operators are now required to restore their sites, once use ceases. A similar requirement should be placed upon industrial operators who would be required to establish sinking funds, to provide for the restoration of land, at the end of its current use. Of course, operators facing bankruptcy are unlikely to use what little value they can salvage from a failing enterprise on reclamation. Multinational corporations, abandoning branch plants, are also likely to attempt to minimize their liabilities and outgoings. Growing industrial dereliction, however, presents an increasingly serious challenge to urban regeneration. A national policy response will be required sooner or later. The proposed urban regeneration agency, English Partnerships, working alongside local authorities, could make a very significant contribution over a five to ten year period if it were adequately funded, perhaps by a small surcharge on the uniform business rate. Powers should be granted to the agency to acquire, at nil value, any land or buildings which become officially derelict. Tenders would then be invited from the public and private sectors, and the land leased to whichever organization would undertake its reclamation at least cost to the agency.

10.6.8 Urban land recycling grant

The case studies showed that the most helpful forms of financial assistance in bringing industrial land forward for development in the Cheshire—Wirral corridor came from derelict land grant and from the European Regional Development Fund. No manufacturing companies in the sample obtained city grant or its immediate predecessors, urban development grant and urban regeneration grant. Indeed, these grants appeared to be of relatively limited importance to industrial development in the Cheshire—Wirral corridor. This experience suggests that, in the long term, the most effective contribution which financial assistance can make towards the renewal of the industrial fabric is to enhance the quality of available sites through encouraging reclamation, servicing and the provision of infrastructure. This could be achieved by widening the scope of derelict land grant, increasing the resources available, and renaming it urban land recycling grant.

Urban land recycling grant would be available on an equal and competitive basis to those in both the public and private sectors able to make best use of it. As with the current derelict land grant, the new grant should be administered by English Partnerships, hopefully with much greater flexibility than the Department of the Environment administered derelict land grant. The grant would be intended to cover any loss involved in reclaiming, servicing, landscaping and providing infrastructure on problematic sites. To encourage demand for such funds, at least in the short term, industrial building allowances could be reintroduced but only on a strictly targeted basis. For instance, if such allowances were made available only on units of under 1000 square metres constructed by manufacturing companies on recycled land, it would both encourage small and medium sized manufacturing companies to invest in their own premises and would provide brownfield sites with a competitive advantage over greenfield sites.

These recommendations would help create a policy response which would seek not only to promote remedial action on a site by site basis, but also to channel resources towards those agencies and institutional arrangements most likely to succeed in making land available for industrial development in the future. The recommendations depend upon much closer cooperation, especially within the public sector, between various development professions. Close working relationships between planning and estate officers (including the possibility of joint project teams), for example, should be promoted within any efficient local authority. Some of the recommendations can be implemented within existing powers and resources. Others will require a redirection of existing resources. However, more extensive recycling of redundant industrial land will occur only when additional powers and resources are made available to create an effective policy response to growing industrial dereliction.

APPENDIX A

INFORMATION SOUGHT FOR THE 120 SAMPLE SITES

- Brief details to identify the site (e.g. size).
- Development status (developed, under construction, undeveloped).
- Existing buildings on site.
- Adverse ground conditions.
- Access to site.
- Services on site.
- Other physical constraints on site.
- Application for government grant aid.
- Planning history 1985—9.
- Development process, including identification of factors most responsible for development delays and of future plans for site.
- Ownership of site including details of transactions, valuations and marketing arrangements.
- Company location decisions and future plans.

APPENDIX B

LIST OF CASE STUDIES

APPENDIX C

INDEX TO KEY FEATURES OF THE CASE STUDIES

The following index is not a comprehensive guide to the case studies. Those highlighted are indicative of key features of the development process. They illustrate aspects of the behaviour of the various agents in the industrial land market; the several development constraints identified; and the mechanisms, strategies and structural factors which facilitate development.

1. THE ROLE AND BEHAVIOUR OF AGENTS IN THE INDUSTRIAL LAND MARKET

2. CONSTRAINTS ON INDUSTRIAL DEVELOPMENT

3. MECHANISMS TO BRING FORWARD SITES FOR DEVELOPMENT

BIBLIOGRAPHY

Adams, C.D. (1987) Industrial Property: The Nature of Demand for Small Premises, 283 *Estates Gazette*, pp. 533-535.

Adams, C.D. (1990) Meeting The Needs of Industry? The Performance of Industrial Land and Property Markets in Inner Manchester and Salford, in Healey, P. and Nabarro, R. (eds) *Land and Property Development in a Changing Context*, Gower, Aldershot.

Adams, C. D. , Baum, A.E. and MacGregor, B. D. (1985) The Influence of Valuation Practices upon the Price of Vacant Inner City Land, *Land Development Studies*, **2**, No. 3, pp. 157-173.

Adams, C. D. , Baum, A.E. and MacGregor, B. D. (1987) Vacant Urban Land: The Planners' Responsibility?, *The Planner,* **73**, No. 6, pp. 31-34.

Adams, C. D., Baum, A.E. and MacGregor B. D. (1988) The Availability of Land for Inner City Development: A Case Study of Inner Manchester, *Urban Studies* , **25**, No. 1, pp. 62-76.

Adams, C.D. and May, H. G. (1991) Active and Passive Behaviour in Land Ownership, *Urban Studies*, **28**, pp. 687-705.

Aims of Industry (1989) *Government v government: The Chester Case*, Aims of Industry, London.

Barrett, S., Stewart, M and Underwood, J. (1978) *The Land Market and the Development Process*, S.A.U.S. Occasional Paper No. 2, University of Bristol.

Breheny, M. and McQuaid, R. (1985) *The M4 Corridor: Patterns and Causes of Growth in High Technology Industries*, Geographical Paper No. 87, Department of Geography, University of Reading.

Burrows, J. W. (1977) *Vacant Urban Land and its Planning Implications*, Unpublished M. Phil. Thesis, University College, London.

Cameron, C. (1973) Intra-urban Location and the New Plant, *Papers of the Regional Science Association*, **31**, pp. 124-143.

Cameron, S and Fleming G. (1990) Implications of Local Property Market Variation for Economic Development Policies: A Case Study of Wakefield, in Healey, P. and Nabarro, R. (eds) *Land and Property Development in a Changing Context*, Gower, Aldershot.

Cameron, G.C., Monk, S. and Pearce, B. J. (1988) *Vacant Urban Land: A Literature Review 1976-86*, Department of the Environment, London.

Castells, M. (1975) *Sociology of Industrial Space*, Anthropos Editions, Paris.

Champion, A. G. and Townsend, A. R. (1990) *Contemporary Britain. A Geographical Perspective,* Edward Arnold, London.

Chandler, J. A. and Lawless, P. (1985) *Local Authorities and the Creation of Employment*, Gower, Aldershot.

Cheshire County Council (1985) *Cheshire County Structure Plan, including the First Alteration, Explanatory Memorandum and Written Statement*, County Planning Department, Cheshire County Council, Chester.

Cheshire County Council (1990) *Cheshire 2001: A Strategy for the Future, County Structure Plan, Submitted Second Alterations Explanatory Memorandum,* Environmental Planning Service, Cheshire County Council, Chester.

Cheshire County Council (1992) *Cheshire Replacement Structure Plan*, Environmental Planning Service, Cheshire County Council, Chester.

Cheshire County Council (1993a) *Land for Jobs No. 3*, Environment Planning Service, Cheshire County Council, Chester.

Cheshire County Council (1993b) *Cheshire 2000: Replacement County Structure Plan*, Environmental Planning Service, Cheshire County Council, Chester.

Chester City Council (1983) *Greater Chester District Plan*, Department of Development and Technical Services, Chester City Council, Chester.

Chester City Council (1985) *Chester Rural Areas Local Plan*, Department of Development and Technical Services, Chester City Council, Chester.

Chester City Council (1988) *Greater Chester Draft Local Plan*, Department of Development and Technical Services, Chester City Council, Chester.

Chisholm, M. and Kivell, P. (1987) *Inner City Waste Land,* Hobart Paper No. 108, Institute of Economic Affairs, London.

Civic Trust (1988) *Urban Waste Land Now,* Civic Trust, London.

Cooke, P. (1989) Locality, Economic Restructuring and World Development, in Cooke, P. (ed.) *Localities*, Unwin Hyman, London.

Coopers and Lybrand Associates and Drivers Jonas (1980) Provision of Small Industrial Premises, Department of Industry, London.

Coopers and Lybrand Associates and Drivers Jonas (1986) *Accommodation Needs of Modern Industry,* Department of the Environment, London.

Crewe & Nantwich Borough Council (1990) *Crewe Local Plan*, Department of Technical Services, Crewe & Nantwich Borough Council, Crewe.

Crewe & Nantwich Borough Council (1990) *Nantwich Local Plan*, Department of Technical Services, Crewe & Nantwich Borough Council, Crewe.

CSW (1989) Consultants Point to Nationwide Industry Shortage, *Chartered Surveyor Weekly*, 27th April 1989, p.17.

Denyer-Green, B.(1989) (3rd edn.) *Compulsory Purchase and Compensation*, *Estates Gazette*, London.

Department of the Environment (1980) *Circular 22/80, Development Control—Policy and Practice*, HMSO, London.

Department of the Environment (1984a) *Circular 14/84, Green Belts*, HMSO, London

Department of the Environment (1984b) *Circular 16/84, Industrial Development*, HMSO, London.

Department of the Environment (1985) *Circular 14/85, Development and Employment*, HMSO, London.

Department of the Environment (1986) *Circular 2/86, Development by Small Businesses,* HMSO, London.

Department of the Environment (1988a) *Planning Policy Guidance Note 4: Industrial and Commercial Development and Small Firms*, HMSO, London.

Department of the Environment (1988b) *Planning Policy Guidance 7: Rural Enterprise and Development*, HMSO, London.

Department of the Environment (1988c) *Planning Policy Guidance 11: Strategic Guidance for Merseyside*, HMSO, London.

Department of the Environment (1988d) *Planning Policy Guidance 2: Green Belts*, HMSO, London.

Department of the Environment (1992a) *Cheshire Replacement Structure Plan: Modifications and Notice of Approval,* Department of the Environment Regional Office, Manchester.

Department of the Environment (1992b) *Proposals to Prevent Land Becoming Derelict,* Department of the Environment, London.

Department of the Environment (1992c) *Planning Policy Guidance 7 (Revised): The Countryside and the Rural Economy*, HMSO, London.

Department of the Environment (1992d) *Planning Policy Guidance 4 (Revised) Industrial and Commercial Development and Small Firms*, HMSO, London

Department of the Environment (1993) *Urban Regeneration Agency. Draft Core Guidance*, Inner Cities Policy Unit, London.

Ellesmere Port & Neston Borough Council (1989) *Ellesmere Port & Neston Draft Local Plan (excluding Mersey Marshes), Background Paper No. 4,* Borough Environment Services Department, Ellesmere Port & Neston Borough Council, Ellesmere Port.

Ellesmere Port & Neston Borough Council (1990) *Ellesmere Port & Neston*

Deposit Local Plan (excluding Mersey Marshes), Borough Environment Services Department, Ellesmere Port & Neston Borough Council, Ellesmere Port.

Evans, A. (1983) The Determination of the Price of Land, *Urban Studies*, **20**, No. 2, pp. 119-129.

Evans, A. (1985) *Urban Economics*, Basil Blackwell, Oxford.

Evans, A. (1986) The Supply of Land: A Pedagogic Comment, *Urban Studies*, **23**, No. 6, pp. 527-530.

Fielding, A.J. (1989) *Inter-Regional Migration: A Study of the South-East of England*, Institute of British Geography, New Series, **14**, pp.24-36

Fothergill , S. and Gudgin, G. (1979) In Defence of Shift-Share, *Urban Studies*, **16**, No. 3, pp. 309-319.

Fothergill, S. and Gudgin, G. (1982) *Unequal Growth: Urban and Regional Employment Change in the U.K.*, Heinemann, London.

Fothergill, S., Kitson, M. and Monk, S. (1983a) *Changes in Industrial Floorspace and Employment in Cities, Towns and Rural Areas*, Industrial Location Research Project, Department of Land Economy, University of Cambridge, Working Paper No. 4.

Fothergill, S., Kitson, M. and Monk, S. (1983b) *Industrial Land Availability in Cities, Towns and Rural Areas*, Industrial Location Research Project, Department of Land Economy, University of Cambridge, Working Paper No. 6.

Fothergill, S., Monk, S. and Perry. M. (1987) *Property and Industrial Development*, Hutchinson, London.

Gloster, M. and Smith, N. (1989) *Inner Cities—A Shortage of Sites*, Royal Institution of Chartered Surveyors, London.

Goodchild, R. and Munton, R. (1985) *Development and the Landowner: An Analysis of the British Experience*, George Allen and Unwin, London.

Gore, A. and Nicholson, D.J. (1985) The Analysis of Public Land Ownership and Development, in Barrett, S. and Healey P. (eds.) *Land Policy: Problems and Alternatives*, Gower, Aldershot.

Gore, T. and Nicholson, D. (1991) Models of the Land-Development Process: A Critical Review, *Environment and Planning A*, **23**, pp. 705-730.

Healey, P. (1986) Emerging Directions for Research on Local Land Use Planning, *Environment and Planning B: Planning and Design*, **13**, pp. 103 - 122.

Healey, P. (1991a) Models of the Development Process: A Review, *Journal of Property Research*, **8**, pp. 219 - 238.

Healey, P. (1991b) Urban Regeneration and the Development Industry,

Regional Studies, **25**, No. 2, pp. 97-110.

Healey, P. and Barrett, S. (1985) Land Policy: Towards a Research Agenda, in Barrett, S. and Healey, P. (eds) *Land Policy: Problems and Alternatives*, Gower, Aldershot.

Healey, P. and Barrett. S.M. (1990) Structure and Agency in Land and Property Development Processes: Some Ideas for Research, *Urban Studies*, **27**, No. 1, pp. 89-104.

Hall, P. (1988) The Industrial Revolution in Reverse?, The Planner, **74**, No. 1, pp. 15-19.

Healey, M. and Clark D. (1984) Industrial Decline and Government Response in the West Midlands: the Case of Coventry, *Regional Studies*, **18**, No. 4, pp. 303-318.

Healey, P., McNamara, P., Elson, M and Doak, A. (1988) *Land Use Planning and the Mediation of Urban Change*, Cambridge University Press, Cambridge.

Healey, P., Davoudi, S., Tavsanoglu, S., O'Toole, M. and Usher, D. (1992) *Rebuilding the City*, E & FN Spon, London.

Henderson, J. and Castells, M. (1987) *Global Restructuring and Territorial Development*, Sage Publications, London.

Henneberry, J. (1988) Conflict in the Industrial Property Market, Town Planning Review, **59**, No. 3, pp. 241-262.

Hobbs, P. (1991) The Response of Local Planning Activity to Economic Change: The Case of Coventry, in Nadin, V. and Doak, J. (eds) *Town Planning Responses to City Change,* Avebury, Aldershot.

Howes, C. (1989) Special Report Land Assembly: Private Sector gets a Boost, *Chartered Surveyor Weekly*, 19 January 1989, pp. 61-63.

INWARD (1988) *Strategic Sites and Premises in North West England*, Unpublished Report, Inward, Chorley.

Isard, W.C. (1956) *Location and Space Economy*, M.I.T. Press, Cambridge, MA.

Keeble, D. (1968) Industrial Decentralization and the Metropolis: the Northwest London Case, *Transactions of the Institute of British Geographers*, **44**, pp. 1-54.

Keeble, D. (1992) Regional Development in the 1990s, in Townroe and Martin (eds.) *The British Isles in Transition*, Regional Studies Association, London.

Keeble, D., Owens, P. and Thompson, C. (1983) The Urban—Rural Manufacturing Shift in the European Community, *Urban Studies*, **20**, No. 4, pp. 405-418.

Lawless, P. (1989) *Britain's Inner Cities*, Paul Chapman Publishing Ltd,

London.

Lloyd, P. E. and Shutt, J. (1983) *Recession and Restructuring in the North-West Region: The Policy Implications of Recent Events*, NWIRU Working Paper No. 13, Department of Geography, University of Manchester.

Losch, A. (1954) *The Economics of Location*, Yale University Press, New Have, CT.

McDermott, P. J. (1973) Spatial Margins and Industrial Location in New Zealand, *New Zealand Geographer*, **29**, pp. 64-74.

MacGregor, B.D., Baum, A.E. , Adams, C.D., Fleming, S.C. and Peterson, J. (1985) *Land Availability for Inner City Development,* Working Paper in Environment Policy No. 8, Department of Land Management and Development, University of Reading.

Markowski, S. (1978) *Study of Vacant Land in Urban Areas*, Centre for Environmental Studies, London.

Massey, D. and Catalano, A. (1978) *Capital and Land: Land Ownership by Capital in Great Britain*, Edward Arnold, London.

Massey, D. and Meegan, R. (1982) *The Anatomy of Job Loss,* Methuen, London.

Mason, C. M. (1989) Explaining Recent Trends in New Firm Formation in the U.K. : Some Evidence from South Hampshire, *Regional Studies*, **23**, No. 4, pp. 331-346.

McNamara, P. F. (1983) Towards a Classification of Land Developers, *Urban Law and Policy*, **6**, pp. 87-94.

McNamara, P. F. and Healey, P. (1984) *The Limitations of Development Control Data in Planning Research,* Town Planning Review **55** (1).

Merseyside County Council (1980) *Merseyside Structure Plan*, County Planning Department, Merseyside County Council, Liverpool.

Merseyside County Council (1983) *Merseyside Green Belt Local Plan*, Department of Development and Planning, Merseyside County Council, Liverpool.

Mills, L. and Young, K. (1986) Local Authorities and Economic Development: A Preliminary Analysis, in Hausner, V.A. (ed.) *Critical Issues in Urban Economic Development*, Clarendon Press, Oxford.

Morgan, E. (1990) A Shortage of Factories, in Healey, P. and Nabarro, R. (eds) *Land and Property Development in a Changing Context*, Gower, Aldershot.

Neutze, M. (1987) The Supply of Land for a Particular Use, *Urban Studies*, **24**, No. 5, pp. 379-388.

Oakey, R.P. and Cooper, S.Y. (1989) High Technology Industry, Agglomeration and the Potential for Peripherally Sited Small Firms,

Regional Studies, **23**, No. 4, pp. 347-360.

O'Farrell, P.N. and Hitchens, D.M.W.N. (1988) Alternative Theories of Small-Firm Growth, *Environment and Planning A.*, **20**, pp. 1365-1383.

Perry, M. (1986) *Small Firms and Economic Development*, Gower, Aldershot.

Price Waterhouse, Casco and Chesterton (1989) *Financing Inner Cities: A Casebook,* Price Waterhouse, London.

Ratcliffe, J. (1978) *Urban Land Administration*, Estates Gazette, London.

Righter, R. (1977) *Save Our Cities*, Calouste Gulbenkian Foundation, London.

Robson, B. (1988) *Those Inner Cities*, Oxford University Press, Oxford.

Roger Tym & Partners (1989) *Greater Manchester Property Market Study*, GMED, Manchester.

Royal Institution of Chartered Surveyors (1978) *Inner City Regeneration*, RICS, London

Sayer, A. (1982) Explaining Manufacturing Shift: a reply to Keeble, *Environment and Planning A*, **14**, pp. 119-125.

Scarrett, D. (ed.) (1988) *Sources of Property Market Information*, Gower, Aldershot.

Scott, A. J. (1982) Locational Patterns and the Dynamics of Industrial Activity in the Modern Metropolis, *Urban Studies*, **19**, No. 2, pp.111-142.

Sellgren, J.M.A. (1991) The Changing Nature of Economic Development Activities: A Longitudinal Analysis of Local Authorities in Great Britain, 1981 to 1987, *Environment and Planning C, Government and Policy*, **9**, pp. 341-362.

Shepherd, J., and Abakuks, A. (1992) *The National Survey of Vacant Land in Urban Areas of England, 1990*, HMSO, London.

Spencer, K. (1987) The West Midlands: An Economy in Crisis, in Hausner, V.A. (ed.) *Urban Economic Change, Five Case Studies,* Clarendon Press, Oxford.

Smith, D. (1966) A Theoretical Framework for Geographical Studies of Industrial Location, *Economic Geography*, **42**, pp. 95-113.

Smith, D. (1981) (2nd Ed) *Industrial Location*, Wiley, London.

Taylor, M. J. (1970) The Location Decisions of Small Firms, *Area*, **2**, pp. 51-54.

Townroe, P. M. (1991) Rationality in Industrial Location Decisions, *Urban Studies*, **28**, No. 3, pp. 383-392.

Townroe, P.M., and Martin R. (1992) *Regional Development in the 1990s. The British Isles in Transition,* Regional Studies Association, London.

Valuation Liaison Group (1985) *Professional Guidance Note No. 2, Section*

2: Disposals of Inner—City Housing Land - Consideration of Special Problems, V.L.G., London.

Watson, G. (1987) *Recycling Disused Industrial Land in the Black Country*, Working Paper No. 97, Department of Town Planning, Oxford Polytechnic.

Watts, H. D. (1987) *Industrial Geography*, Longman, Harlow.

Weber, A. (1929) *Theory of the Location of Industries*, Chicago University Press, Chicago.

Wiltshaw, D. G. (1985) The Supply of Land, *Urban Studies*, **22**, No. 1, pp. 49-56.

Wiltshaw, D. G. (1988) Pedagogic Comment and the Supply of Land for a Particular Use, *Urban Studies*, **25**, No. 5, pp. 439-447.

Wirral Metropolitan Borough Council (1990a) *Unitary Development Plan: Issues and Draft Part One Policies*, Department of Planning and Economic Development, Wirral Metropolitan Borough Council, Wallasey.

Wirral Metropolitan Borough Council (1990b) *Unitary Development Plan: Issues and Draft Part One Policies: Appendices*, Department of Planning and Economic Development, Wirral Metropolitan Borough Council, Wallasey.

Wirral Metropolitan Borough Council (1992) *Unitary Development Plan. Draft for Public Consultation,* Wirral Metropolitan Borough Council, Wallasey.

SUBJECT INDEX

AUTHOR INDEX